A POIESIS OF THE CREATIVE COSMOS:

Celebrating Her within PaGaian Sacred Ceremony

Glenys Livingstone Ph.D.

Girl God Books

Cover art by Julie Cunningham

ISBN: 978-82-93725-44-2
Black and White Edition

Girl God Books are also available at discounts for retail or wholesale bulk purchases.
For details, contact support@girlgod.org.

You can find a resource page with the videos mentioned in this book – along with some bonus features – at https://thegirlgod.com/pagaianresources.php.

A Poiesis of the Creative Cosmos Book Endorsements

"This is a deeply spiritual work, beautifully articulating the sacrality and the profundity of our lives and of the cosmos. Glenys Livingstone writes of communions, celebrations, wonderfully creative ceremonies, the continuity of life: our turns around the Wheel of the Year."
-Miriam Robbins Dexter, Author of *Whence the Goddesses: a Source Book* and *Sacred Display: Divine and Magical Female Figures of Eurasia*

"This book is a blessing! Bestowing knowledge, wisdom, and beauty, it presents new possibilities for a greater felt connection with the cosmos. The ceremonies for a PaGaian Wheel of the Year will surely inspire spiritual creativity."
-Charlene Spretnak, Author of *Lost Goddesses of Early Greece*

"For years and years, we Goddess-folk have chanted, sung and whispered--the Great Mother is returning! Now, in these challenging times, we are blessed with a deeply personal, profoundly universal book from the inimitable Glenys Livingstone. Buy this book, sisters. Read this book, brothers. Live this book, children of Our Sacred Mother." -H. Byron Ballard, Witch, Priestess, Author....

"In these times we need to remember our cosmic origins, the organic power we have within us collectively and personally. Glenys' book offers inspiration for such a journey which is both in and out at the same time: it presents an action that may be entered into to enable our atrophied sensitivities to our Earth, our place of being. The process described in this book helps us remember how old we really are, and also how young and vibrant we may be. As Joanna Macy says: it's time for us to act our age."
-John Seed, deep ecology activist, co-author (with Joanna Macy, Pat Fleming, and Arne Naess) of *Thinking Like a Mountain: Toward a Council of All Beings.*

"In order to enter *A Poiesis of the Creative Cosmos*, begin the night before. Go out under the stars and let their beauty soak into you. Bring to mind the 320 million Earth-like planets sprinkled throughout the Milky Way galaxy and imagine there are other beings out there who are, *simultaneously*, trying to imagine us. As you climb into bed, reflect on the stunning truth coming from both the ancient ones and the contemporary scientists, that you are at the center of the universe, at the very center along with everyone else. After a deep sleep, you will be ready to enter the PaGaian wheel of ceremony which has taken Glenys an entire lifetime to create. She has had to struggle out of the unconscious arrogance and profound ignorance of industrial society in their dismissal of the wisdom embedded in Goddess-based cultures. Working with other poets of the cosmos, she has constructed a spiritual practice that enables us to root ourselves in place and celebrate the sacred annual journey about the sun. By releasing ourselves into the wisdom of her ceremonies, we find our creative energies reinvigorated as we set forth to give poetic expression to the divine presence in which we live. Glenys Livingstone, the author of this cosmological poetry, is a planetary treasure."
-Brian Thomas Swimme, Author of *Cosmogenesis*

"Poiesis, a Greek word meaning "the activity in which a person brings something into being that did not exist before" is perhaps the most elegant word to describe the act of creating ceremony for understanding and connection. This book encourages authenticity in creating meaning-full seasonal ceremony in realationship to land and kinship, in this time and space, right here and now.

As we consciously mark special days through the year, whether they be anniversaries, seasonal shifts or life's transitions, and link those days to stories that are bigger than our own experience, we open more to find deeper understanding and expression. For me, Glenys' work is a supportive reminder that indeed, Creating Ceremony is a key piece in the journey of Self."
-Hollie Bakerboljkovac, Founder of Institute for Self Crafting

Dedication

I thank the Mothers and the Grandmothers of this whole Land, particularly those of my homelands, Wakka Wakka country, and specifically those of the place of the Blue Mountains Australia, Gundungurra and Dharug country, where I learned so much of this; and the Mothers and the Grandmothers of the ages, of many times and places, who have spoken to me in some deep way, beyond my knowing.

The Land calls us Home.
We are this Earth, this Cosmos –
Mother will speak with us …
She is in our every cell.

CONTENTS

List of Photos

PREFACE

The term 'PaGaian,' which became part of the title of my work, was conceived in at least two places on the Planet and in opposite hemispheres within a year of each other, in the early years of the twenty-first century, without either inventor being aware of the other's new expression. It was some time before they found each other … one party in Australia, myself having published a book with PaGaian in the title; and the other party, Rob Blake in the UK having registered the domain *pagaian.org* as the term "pagaian" seemed to him to express a cosmology constellating in his mind. The term *PaGaian* as a naming of my work and practice, was conceived by my partner Robert (Taffy) Seaborne in late 2003, enabling the book to manifest: heretofore the body of work that I had developed took six lines to name (as it did in its doctoral form).

This reaching for a new word, such as "PaGaian/pagaian," was the reaching for a *language,* which is a *power*, to bring together an Earth-based – 'Pagan' – spiritual practice indigenous to Western Europe, with recent Western scientific understandings of the planet as a whole living organism – 'Gaian' as it has been named,[1] and which by its name acknowledged resonance with ancient Mother Goddess understandings of our Habitat, as an alive sentient being. So, the term 'PaGaian' splices together Pagan and Gaian, and it may express a new autochthonic/native context in which humans find themselves: that is, the term may express for some (as it did and does) an indigeneity, a nativity, in these times, of belonging to this Earth, this Cosmos. For myself, the new expression consciously included and centralised female metaphor for sacred practice: that is, practice of relationship with the sacred whole in which we are, and whom I desired to call Mother, and imagine as the Great She.

Human language has been described as "fire in the mind,"[2] based on the idea that perhaps it developed dramatically when early

---

[1] See James Lovelock and Lynn Margulis, the "Gaia Theory."

[2] Brian Swimme, *Canticle to the Cosmos* (CA: Tides Foundation, 1990), video 9.

humans (likely homo erectus)[3] tamed fire and sat around the fire at night, reflecting on their experiences and telling stories: those gifted with language could affect the group powerfully. As Brian Swimme puts it: "a new selection pressure was brought forth – the pressure of linguistic competence:"[4] that is, language acted as a part of the biological shaping power of natural selection. And today this selection pressure is shaping the entire planet. Human language, in its many modalities, now determines and sculpts the biosphere, the lithosphere, the atmosphere and the hydrosphere – by the stories we tell, the structures we put in place.

Amongst Celtic peoples, the capacity to speak poetically was a divine attribute, regarded as a transformative power of the Deity, who was named by those peoples as the Great Goddess Brigid: She was a Poet, a Matron of Poetry (along with her capacities of smithcraft and healing). And at Delphi in Greece, the oracular priestesses delivered their prophecies in poetic form: Phemonoe invented the poetic meter, the hexameter. And from Sumer, humans have the first Western written records of literature, which is poetry written by the Sumerian High Priestess Enheduanna, in approximately 2300 B.C.E. Poetry has been recognised as a powerful modality: Barbara Mor and Monica Sjöö described "poetic thinking" as a *wholistic* mode, wherein "paradox and ambiguity … can be felt and synthesized. The most ancient becomes the most modern; for in the holographic universe, each 'subjective' part contains the 'objective' whole, and chronological time is just one aspect of a simultaneous universe."[5]

---

[3] See Appendix C, *(31).

[4] Brian Swimme, *Canticle to the Cosmos Study Guide* (Boulder CO: Sounds True Audio, 1990), 36, referring to DVD 9, "Fire in the Mind."

[5] Monica Sjöö and Barbara Mor, *The Great Cosmic Mother: Rediscovering the Religion of the Earth* (San Francisco: Harper and Row, 1987), 41.

Poetry could be described as an "Earth-centred language:"[6] it has the capacity to hold multivalent aspects of reality, to open to subjective depths, to allow qualitative differences in understanding, hence it is especially suited to expressing and bringing together a multitude of beings. Cosmologist and evolutionary philosopher Brian Swimme and the late cultural historian/geologian Thomas Berry have called for such a language – the kind of language "until now enjoyed only by our poets and mystics" that may express the "highly differentiated unity,"[7] the organic reality such as Earth is, and such as "Gaia" was understood of old, and in recent scientific theory: that is, as a *highly differentiated unity*, which any expression must aim to emulate.

I have always understood PaGaian Cosmology as Poetry:[8] it is not a 'discourse' or a theory, or a 'study' of something as a theology is, or even as a thealogy may be. It is a *speaking with* our Place, this Habitat, which is understood to be alive and responsive, and deeply complex: how else may we speak with our dynamic Place of Being, who is always much more than we can imagine? The ceremonial celebration of the complete cycle of Seasonal ceremonies, wherever one is on our Planet, and in all the diverse possibilities, may be experienced and recognised as a Poiesis: that is, the intention is to *make* a world, to participate in "an action that transforms and continues the world"[9] … the sacred ceremonies when engaged in fully, are a method of action. They may serve as a catalyst for changing of mind, for personal and cultural change.

---

[6] A term used by Brian Swimme and Thomas Berry, *The Universe Story: From the Primordial Flaring Forth to the Ecozoic Era* (New York: HarperCollins, 1992), 258.

[7] As Earth is described by Swimme and Berry, *The Universe Story*, 259, in the context of calling for "a multivalent language much richer in its symbolic and poetic qualities."

[8] Glenys Livingstone, *PaGaian Cosmology: Re-inventing Earth-based Goddess Religion* (Nebraska: iUniverse, 2005), 42-45.

[9] Wikipedia definition of Poiesis: http://en.wikipedia.org/wiki/Poiesis.

PaGaian Cosmology is primarily an *action* of sacred practice: an art form of ritual/ceremony,[10] which is consistently practised over the full year – the full orbit of Earth around Sun, and which may re-create a *sense* of *sacred* space-time of this everyday journey, which all on the Planet make whether conscious or not. The practice is a re-creating – a *re-creation*, of what I name as Gaia's Womb: a name for the whole sacred site in which we live, and the sensation/ embodiment of which is created with the practice of ceremony of Earth-Sun transitions, the Seasonal Moments, however they manifest in your place.[11] It is the regular conversation throughout the *whole* annual cycle – the sacred gestalt – that creates the womb, the space of integral relationship with Source of Being … whom one may understand as the Great She, birthing all, other and self in every moment.

The sacred site thus created is a space that nurtures the *sense* of the continuum in which we are immersed. Many indigenous cultures still have this sacred relational sense of the world that is nurtured by ceremonies; and many of a variety of cultures in these times of great change seek such a relational sense – and who may identify as being in "recovery from Western civilization."[12] I have been engaged for decades now, in re-turning to my indigenous religious heritage of Western Europe, re-creating, and re-inventing a ceremonial practice that celebrates the sacred journey around Sun: it

---

[10] Kathy Jones, Priestess of Avalon, distinguishes ceremony from ritual; see *Priestess of Avalon, Priestess of the Goddess* (Glastonbury: Ariadne Publications, 2006), 319. I also explain the distinction in Chapter 3.

[11] It is my experience that the created ceremonies do not need to be in the same exact location throughout the year, though that is good when possible: the ceremonies themselves are received into the sentient Cosmos and part of one's own sentience, and if regularly practiced – religiously practiced, the sacred site, Gaia's Womb, may be virtual (intra/meta-physical), as much as it is physical.

[12] A term used by Chellis Glendinning in the title of her book: *My Name is Chellis and I'm in Recovery from Western Civilization* (Boston: Shambhala Publications, 1994).

has been an intuitive, organic process synthesizing bits that I have learned from good teachers and scholars, and bits that have just shown up within dreams and imagination, as well as academic research. It has been a shamanic journey: that is, I have relied on my direct lived experience for an understanding of the sacred, as opposed to relying on an external authority, external imposed symbol, story or image. It has not been a *pre*-scriptive journey: I have scripted it myself, *self*-scribed it, and in cahoots with the many who participated in the storytelling circles, ceremonies and classes over decades. The pathway was and is made in the walking. It is part of a new fabric of understanding – created by new texts and contexts, both personal and communal – that have been emerging in recent decades, and continue so, at awesome speed in our times.

In the telling of the Universe story, Brian Swimme and Thomas Berry have said: "Cosmology aims at articulating the story of the universe so that humans can enter fruitfully into the web of relationships within the universe."[13] ... and in these times when "the role of the human in this web of relationships is changing so radically," a new language is called for. They say:

> To articulate anew our orientation in the universe requires the use of a language that does not yet exist, for each extant language harbors its own attitudes, its own assumptions, its own cosmology. Thus to articulate anew the story of our relationships in the world means to use the words of one of the modern languages that implicitly, and to varying degrees obscures or even denies the reality of these emerging relationships. Any cosmology whose language can be completely understood by using one of the standard dictionaries belongs to a former era.[14]

By way of example – and relevant to PaGaian Cosmology, and the growing population of paganistic practitioners of many varieties: a "sabbat" is defined in the Webster's Dictionary as "a midnight

---

13 Swimme and Berry, *The Universe Story*, 23.
14 Ibid. 24.

assembly of witches and sorcerers held in medieval and Renaissance times at intervals to renew allegiance to the devil … and to celebrate rites (as the Black Mass) and orgies."[15] In light of the current situation where "sabbats" are Earth holy days recognised and celebrated by a variety of people, such a definition is mostly simply amusing, and indicative of a particular ethnocentric mindframe. Another example is the defining of "Halloween" exclusively (as the same text does) as "the evening preceding All Saint's Day: the evening of October 31 often devoted by young people to merrymaking … and playing pranks, sometimes involving petty damage to property,"[16] without any recognition of its sacred indigenous roots: such a definition is again indicative of a bias that is no longer acceptable in the global context that now recognises the historical era of colonization of many Indigenous peoples including those of Western Europe.

To participate in the re-creating of poetic language to name and celebrate the Cosmos is to participate in expressing a new story of the universe which we inhabit; with our words we may spell a new cosmology.[17] Such an enterprise "could hardly be more traditional" as Swimme and Berry have noted, "for relationships are regularly created, explored, developed, ended, and reinvented at every level of being."[18] Or as Starhawk has expressed it:

> She changes everything She touches, and
> Everything She touches, changes.[19]

---

[15] Webster's Dictionary, 1994.

[16] Ibid. 1023.

[17] Thomas Berry read the draft of my book *PaGaian Cosmology* and gave praise that he was "impressed your work and with the vision, courage, and ardor with which you have followed through on a major aspect of the Great Work."

[18] Swimme and Berry, *The Universe Story*, 22.

[19] Part of the Kore chant, Starhawk, *The Spiral Dance: A Rebirth of the Ancient Religion of the Great Goddess* (New York: Harper and Row, 1999), 115.

(In regard to *PaGaian* cosmology: it is not a 'theism' of any kind – not an 'a-theism' nor a 'pan-theism' or a 'panen-theism': nor do I describe it as a 'thealogy,' though some may. It is about a Place – this Cosmos, this Earth – not a Deity. I prefer the term "Cosmology": it is a study of, or engagement with, our Place, which is dynamic, a Verb, not a Noun – it is an Event. I understand myself as a student of the Poetry of the Universe. I think that 'theology' was meant to be poetry: that is, some of its writers understood it was metaphor … what else could it be as it reached to articulate the Great Mystery of Being? But mostly what it became was the description of a dead butterfly pinned in a glass case, not a butterfly that is alive and flitting about the garden – in the act of being. This Place, this Cosmos, in which Earth is, in which we are, may itself be conceived of as deity – or at least as 'source' of being, however one may choose to express it: and all attempts to describe this reality may be understood to be metaphor … metaphor is all we have for an alive, dynamic, diverse reality.)

The point of celebrating Seasonal Moments, is not only the alignment of food production with place, but also the alignment of story with place – conversation with the awesome place in which we find ourselves: its terror and its beauty. Having grown up in the Southern Hemisphere with a Northern Hemisphere and Christian story of place – and one that used exclusively male metaphors, I knew a profound alienation from my place – that was personal, communal and ecological. The re-membering and creating of a Poetry that could express relationship with my Place – my self as *a* Place, and indeed as a *Place* (that is, of substance) and belonging in a *Place* – became an *essential* quest. It was a quest for new language in my heart and on my lips, to express *sacred relationship* with my place … and in my context, it did not yet exist. It is still only beginning.

A new language enabled the ceremonial practice of that sacred relationship, and in that process I learned so much: on Her lap, She taught me and others. It is a self-knowledge in its layered and complex dimensions, as our Place of Being is … the self who is particular, the self who is deeply related, and the self who directly participates in the sentience of the creative Cosmos. It is a self who is founded in *where* we are: regional, Earth and Cosmos – inseparable.

A practice of ceremony that celebrates the whole cycle of EarthGaia's sacred journey around Sun – *Where* we are – may teach a person, grow a person, within the dimensions of *real* time and space[20] – the place from which *true* action may arise.

The following chapters are a documentation of my work of PaGaian Cosmology as it has developed to this point in time, since the publication of my book *PaGaian Cosmology: Re-inventing Earth-based Goddess Religion*, yet also including much that went before, which has been part of the year-long course I have taught, both on the ground in my sacred space and on-line. There may be some minimal repetition in the course of the whole book, for the purpose of clarity in different contexts and also enabling each chapter to stand alone.

---

20 Ecofeminist philosopher Charlene Spretnak also names the foundation of existence as the "Real," which is different from popular notions and clichés about "the real world": see T*he Resurgence of the Real: Body, Nature and Place in a Hypermodern World* (New York: Routledge, 1999*)*.

Introduction

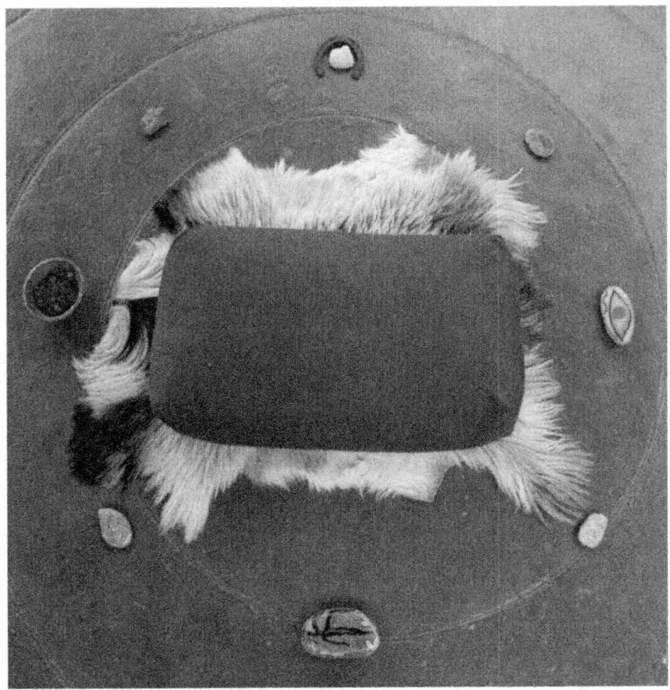

[Figure 1] Meditation cushion in circle of stones

(My ancestors built great circles of stones that represented their perception of real time and space, and enabled them to tell time: the stone circles were cosmic calendars.[21] They went to great lengths and detail to get it right. It was obviously very important to them to have the stones of a particular kind, in the right positions according to position of the Sun at different times of the year, and then to celebrate ceremony within it.)

I have for decades had a much smaller circle of stones assembled. I have regarded this small circle of stones as a medicine wheel. It is a portable collection, that I can spread out in my living space, or let sit in

---

[21] See Martin Brennan, *The Stones of Time: Calendars, Sundials, and Stone Chambers of Ancient Ireland* (Rochester Vermont, Inner Traditions, 1994).

a small circle on an altar, with a candle/candles in the middle. Each stone (or objects, as some are) represents a particular Seasonal Moment/transition and is placed in the corresponding direction. The small circle of eight stones represents the flow of the Solstices and Equinoxes and the cross-quarter Moments in between: that is, it represents the "Wheel of the Year" as it is commonly known in Pagan traditions.

I have found this assembled circle to have been an important presence. It makes the year, my everyday sacred journey of Earth around Sun, tangible and visible as a circle, and has been a method of changing my mind, as I am placed in real space and time. My stone wheel has been a method of bringing me home to my indigenous sense of being. Each stone/object of my small wheel may be understood to represent a "moment of grace," as Thomas Berry named the seasonal transitions – each is a threshold to the Centre, wherein I may now sit: I sense it as a powerful point. As I sit on the floor in the centre of my small circle of stones, I reflect on its significance as I have come to know the Seasonal transitions that it marks, over decades of celebrating them. I sense the aesthetics and poetry of each.

I facilitated and was part of the celebration and contemplation of these Moments in my region for decades. It was always an open group that gathered, and so its participants changed over the years but it remained in form, like a live body which it was: a ceremonial body that conversed with the sacred Cosmos in my place. We spoke a year-long story and poetry of never-ending renewal – of the unfolding self, Earth and Cosmos. We danced and chanted our relationship with the Mother, opened ourselves to Her Creativity, and conversed with Her by this method. All participants in their own way within these ceremonies made *meaning* of their lives – which is what I understand *relationship* to be, in this context of Earth and Sun, our *Place* and Home in the Cosmos: that is, existence is innately meaningful when a being knows Who one is and Where one is. Barbara Walker notes that religions based on the Mother are free of the "neurotic" quest for indefinable meaning in life as such religions "never assumed that life would be required to justify itself."[22]

[22] Barbara Walker, *The Woman's Encyclopedia of Myths and Secrets* (San Francisco: Harper and Row, 1983), 693.

[Figure 2] Meditation cushion in circle of decorated wheel of stones

I face the North stone, which in my hemisphere is where I place the Summer Solstice. From behind me and to my right is the light part of the cycle – representing manifest form, all that we see and touch. From behind me and to my left is the dark part of the cycle – representing the manifesting, the reality beneath the visible, which includes the non-visible. The Centre wherein I sit, represents the present. The wheel of stones has offered to me a way of experiencing the present as "presence," as it recalls in an instant that,

> That which has been and that which is to come are not elsewhere – they are not autonomous dimensions independent of the encompassing present in which we dwell. They are, rather, the very depths of this living place – the hidden depth of its distances and the concealed depth on which we stand.[23]

---

[23] David Abram, *The Spell of the Sensuous* (New York: Vintage Books, 1997), 216.

This wheel of stones, which captures the Wheel of the Year in essence, locates me in the deep present, wherein the past and the future are contained – both always gestating in the dark, through the gateways. And all this has been continually enacted and expressed in the ceremonies of the Wheel of the Year, as the open, yet formal group has done them, mostly in the place of Blue Mountains, Australia.

Over the years of practice of ritually celebrating these eight Seasonal Moments – Earth's whole annual journey around Sun, I have been held in this creative story, this *Story of Creativity* as it may be written – it is a sacred story. Her pattern of Creativity can be identified at all levels of reality – manifesting in seasonal cycles, moon cycles, body cycles – and to be aligned with it aligns a person's core with the Creative Mother Universe. I have identified the placing of one's self within this wheel through ceremonial practice of the whole year of creativity, as the placing of one's self in Her Womb – Gaia's Womb, a Place of Creativity. All that is necessary for Creativity is present in this Place. All may come forth from here/Here – and so it does, and so it has, and so it will.

[Figure 3] PaGaian Cosmology altar/mandala:
a "Womb of Gaia" map

I have summarised and imaged a PaGaian cosmology with the pattern and motifs of this particular altar/mandala in the photo (Figure 3), though it will always vary somewhat according to place (hemisphere and region) and personal expression. I have often entered into this altar/mandala for presentations and introductions in the following way:

TO ORIENT YOU FIRST
I have Water in East, Fire in North, Earth in West and Air in South: which seems appropriate to my place on the Planet – Southern Hemisphere, East Coast Australia. In the Northern Hemisphere Fire would be in the South, and traditionally Air in the East, Water in the West, and Earth in the North … but the elements may be arranged as suitable for any particular place on the Planet.

It is a three-layered arrangement. At the edges are the four elements in four directions – Water, Fire, Earth, Air. These are things I can identify as held in common to all being on this planet, and their sensate presence can be ***felt*** *within any being.*

STEPPING AROUND THE CIRCLE
TO THE EAST
You may feel the Water in you, this primordial ocean recycled many times – its moistness, taste it in you, this ancient Water.

TO THE NORTH
You may feel the Fire in you, this ancient heat passed on – its warmth. Feel its warmth in you, this ancient Fire in you.

TO THE WEST
You may feel the Earth in you, this geological formation– its weight, feel your weight – this Earth in you, this ancient Earth in you.

TO THE SOUTH
You may feel the Air in you – its expansion – this ancient river that all have breathed. Feel this Air expand in you.

Within that boundary is a circle formed by eight objects, representing the eight seasonal points that I and any others have celebrated in recognition of the regional phases of our planet's relationship with our source of energy and life, the Sun. It thus represents our everyday sacred Journey, as Earth around Sun.

STEPPING BACK SLIGHTLY TO THE SOUTH WEST
Deep Autumn/Samhain … death of the old, conception of the new
STEPPING TO SOUTH
Winter Solstice/Yule … Birth of the new, Birth of All – Origins
STEPPING TO SOUTH EAST
Early Spring/Imbolc…nurturance of & dedication to the unique self
STEPPING TO EAST
Spring Equinox/Eostar … the Joy and Power of Being
STEPPING TO NORTH EAST
High Spring/Beltaine … the fertility, Desire and Dance of Life
STEPPING TO NORTH
Summer Solstice/Litha … the Fullness and Wholeness of Being
STEPPING TO NORTH WEST
Lammas/Late Summer … the Harvest of Life – the Sacred Consuming
STEPPING TO WEST
Autumn Equinox/Mabon … the Grief and Power of Loss
STEPPING TO SOUTH WEST
And back to Samhain/Deep Autumn – deep transformation … the circle begins again

The centre of the arrangement is formed by three candles and associated objects which represent the Triple Spiral, the three faces of the Metaphor for the Creative Dynamic that I perceive continually unfolds it all.

This Creative Dynamic …
WEST SOUTH-WEST POSITION
She is the Old One – Who creates the Space to Be – the deep sentience within all … placed in the darkest quarter of the year.
WALKING THE CIRCLE TO EAST SOUTH-EAST POSITION
She is the Young One/Maiden – Who is the Urge to Be – the new continually emerging … differentiated being: placed with the growing light.
WALKING THE CIRCLE TO NORTH NORTH-WEST POSITION
She is the Mother – Who is the Place of Being – the relational Web of the Present … placed with the peaking of light and the turning into the dark, a dynamic place of

flux.

We may remember, **you** *may remember, something of what it has taken to unfold this present Moment and Place – wherever you are in this Moment – the work/creativity of the ancestors, of the ages (from many lands) and those of recent times; the first people of the Land you are on – particularly the local ones of your region, the name of their tribe/group; the creatures and plants of all times and places, the Land Herself. We may be thankful to All – present in the Depth of this Moment and Place.*

LIGHT THE THREE CANDLES

[Figure 4] The Three Candles

And so it is, as I begin to tell you something of this story
– this Poiesis of the Creative Cosmos,
which you may make your own Poiesis,
in your own unique way – for self,
and perhaps with a community – with others,
and always in direct relationship with your place in this Cosmos.

Re-Storying Goddess/Dea

At the heart of PaGaian Cosmology is the re-storying and expression of Goddess metaphor for the sacred: it was She who called me – into Her, to learn of Her, to find a way to speak of Her. This cosmology is originally a study and embodiment of Goddess in three qualities – often known commonly as "Virgin, Mother and Crone," but globally She has been named and praised in various terms: such as possessing the three qualities of 'preserver/protector,' 'creative power,' and 'destructive power' (Kali in India); or in other ancient depictions the three qualities are represented perhaps with grain, sword and snake (Hecate in Greece); perhaps with grain, throne and scorpion (Anatha of Egypt); perhaps as poet, physician and smith-artisan as in the case of Celtic Brigid. Sometimes She has been represented as three matrons (Germany and Italy). In East Asia, there are many triplicities and triads: in Korea Mago, the Creatrix, is identified with Samsin (Triad Deity) and also Goma is referred to as one of the "Three Sages."[24] In South America, the Goddess Chia is known as a triple goddess.

In our times *She* and Her multivalent dimensions have rarely been understood, and frequently Her triplicity has been re-configured as three sages or kings; and in some religions She has been replaced with an all-male trinity. Yet many continued to seek Her.

---

[24] Helen Hye-Sook Hwang is doing much original research of the Magoist Cosmogony. According to Hwang, the number three is an epitome of musically charged nine numbers (3x3). Thus, the triad symbol encoded the metamorphic force of the universe or matriverse, a term coined by Hwang. See *The Mago Way: Re-discovering Mago, the Great Goddess from East Asia, Volume 1* (Mago Books, 2015), particularly chapters 6, 7, and 8. See also "Unveiling an Ancient Sill Korean Testimony to the Mother World: An Introductory Discussion of the Budoji (Epic of the Emblem Capital City), the Principal Text of Magoism" in *S/HE: An International Journal of Goddess Studies* 1 (2, 2022): 22-50. For the three sages, see "Goma, the Shaman Ruler of Old Magoist East Asia/Korea, and Her Mythology", in *Goddesses in Myth, History, and Culture* (Lytle Creek: Mago Books, 2018), 257-270.

The Form and the Shape that they sought
was not in any Atlas.
Her gaps had been covered up,
Her hollows filled in,
Her name blanked out.
She lay buried beneath things, silent,
but with a detectable visceral pulse.[25]

Re-Storying Goddess for me has meant re-storing a sense of *She* to the Cosmos, restoring female sacrality – to the small particular self to begin with, and to other, and to all-that-is: re-storing *Her* as Language, as sacred language – image and word – for the Creative Dynamic that unfolds the Universe. Why not? She has been absent for so long. And what might happen then? What difference might it make to the world we live in?

The primary Place that She, Mother-Universe, may be sensed as present, is in one's own bodymind, the breath, the whole phenomena of its ebbs and flows: and the breath may be the primary place for contemplating Her three aspects of waxing, peaking and waning – out of the void and back into it, over and over. A lot of inhabitants of Western industrialised culture have been turned into outsiders in our own land[26] – and I mean primarily the 'land' of our own bodyminds, but it is also and inseparably true of the Earth land/country in which we dwell. In the patriarchal context, which dominates most present global cultures, this is particularly true for women. We are often (or have been) outsiders in our own land – **the** Land of our own bodyminds: the female in particular, by and large, at some point in the story, lost native title to her land. Most of us – female and male and all variations, have learned well how to think from outside ourselves: and with a consciousness that treats ourselves personally and thus usually also others, as less than worthy of reverence. Women particularly have been and are a colonized people: that is, female embodiment has become the unknown, unspoken, clouded continent in most present global cultures. An overwhelming majority of females on

---

25 Poem by Glenys Livingstone, 1998.

26 I will speak primarily out of my own cultural context, though dualism and alienation from bodymind may be identified cross-culturally.

the planet are engaged in a daily claiming or re-claiming of our native entitlement to our land, our bodymind: the details vary from culture to culture, and situation, and is more extreme/blatant in some contexts, but the overall effect is common, and ultimately shared.

As Elizabeth Cady Stanton said, in the late nineteenth century, it is not female biology that has betrayed us, but the beliefs and the stories we have about ourselves. Most of the religious stories that most on the globe grew up with at this time, did not have the female in *care-ful* mind. Most world religions at this point in time still specifically story the female as problematic to creation or enlightenment or whatever, or at least secondary: though some are clever enough to attempt to disguise it. The term 'God' still commonly invokes the 'Face of Ultimate Reality,' the Absolute, and the term 'Goddess' still overwhelmingly invokes a mere mythological entity.

'Goddess,' or 'God,' is metaphor, a poetic image – suggesting a likeness of femaleness or maleness in the Sacred/Deity. According to the Webster's Dictionary, a metaphor is a word or phrase used to suggest a likeness. 'Goddess' is a figure of speech suggesting a likeness of femaleness in the Deity – few would argue with that: though many do argue that 'God' does not suggest a likeness of maleness in the Deity, that it/He is neutral or may represent both, whereas Goddess may not. Some cultures don't need the word 'God' nor do some need the word 'Goddess' because female sacrality is not a problem – femaleness in such cultures, is understood to embody transformatory powers[27] – and replicates the very nature/essence of the Cosmos: birthing, lactating, conceiving, gestating are understood as cosmic transformatory, regenerative powers. And all genders/sexes in such cultures find purpose in supporting regenerative capacities.

The texts we choose for our lives create the texture, the context – when we choose a story for our lives, or accept a *pre-scribed* one, it lives us. There is a cosmology in our everyday speech and action. Re-storying Goddess, and celebrating Her in Seasonal Moments however they manifest in your region, may participate in the process of *scribing* one's

[27] This is a term used by Melissa Raphael, in *Thealogy and Embodiment: the Post-Patriarchal Reconstruction of Female Sexuality* (Sheffield: Sheffield Press, 1996).

self, authoring one's self, at the deepest level, and re-storying the regenerative Universe as *She*.

When I speak of 'Goddess,' I mean Her as a totality – not a 'Feminine' PART of the Sacred. 'Goddess' may be metaphor for Ultimate Creativity – the Sacred Cosmos … and in three qualities of Her Unfolding, in Her Cosmogenesis: that is,

- as ever-new differentiated being
- in infinite full communion/relatedness
- and with constant transformation within Her sentient Self.

... three qualities that may be sensed multivalently, a triplicity implicit within the creative Cosmos.[28] Brian Swimme named the three qualities of Cosmogenesis,[29] as "cosmic grammar:"[30] that is, as I might say it, "this is how She speaks." The whole Cosmos is Her expression, Her unfolding Body, in never-ending renewal.

She may be re-storied to that integrity in our hearts and minds. Then one *Enters* (becomes en-tranced with) the richness and magic of this Cosmic Dynamic of Creativity, who is embodied and embedded in all being. She is the seamless sacred Matter, Mother, Materia, Madre, whom we are.

The Triple Spiral

At the centre of the altar/mandala which tangibly summarises this PaGaian cosmology, is an image of the Triple Spiral from Brú na Bóinne ("Newgrange" as it is commonly known) in Ireland.

---

28 Caitlin Matthews refers to this "innate triplicity" in "The Triskele of Energy," *The Celtic Spirit* (London: Hodder and Stoughton, 2000), 366, and to its essential nature in "The Threefold," *The Celtic Spirit,* 138.

29 Swimme and Berry describe these qualities in *The Universe Story*, 66-79. I will develop these qualities later.

30 Swimme, *Canticle to the Cosmos*, program 4, "The Fundamental Order of the Universe."

[Figure 5] Triple Spiral of Brú na Bóinne

This Triple Spiral is just one motif amongst a whole collection of art that represents the first major Western European art tradition since the Ice Age.[31] The significance of this collection has been largely unrecognised, because its context of megalithic mounds spread over an area, has not been understood. This complex art collection is engraved on stones throughout the large stone structures, that are dated between 3200 and 3700 B.C.E. – which places these mounds among the world's oldest remaining buildings.[32] The Triple Spiral then is an ancient highly abstract visual design, left by the ancestors of this place, in a context whose meaning is still being unravelled and contemplated.

The story of the un-covering of the Triple Spiral at 'Newgrange' is one that I can identify with. The indigenous name of the place is Brú

---

[31] Brennan, *The Stones of Time*, 37.

[32] Ibid., 7.

na Bóinne – 'Bóinne' being derivative of Boann, Goddess Creator of the River Boyne. By 1142 C.E., the humans inhabiting this place had long forgotten its numinosity, and the place was part of farms known as 'granges,' and by 1378 the mound that is home to the Triple Spiral "had been completely stripped of its former identity and was called merely 'the new grange.'"[33] The process of remembering and unfolding the significance of the place in the context of the many other nearby sites similar to it, over the hundreds of years since then, has been fraught with complexities. The motif itself – the Triple Spiral – has only received the light of the Winter Solstice again, as it was evidently designed to do, since the 1960's, after a millennium or two of being hidden away. Since then, there has been a slow dawning in the minds of some contemplating it, that one of the valencies of this motif is that it may represent the indigenous Goddess, the Land Herself, as She was known there in Her three aspects – Ériu, Fódla and Banba[34] – this being the Sacred metaphor for the great cosmic energies. The megalithic mounds of which Brú na Bóinne is a part, are oriented towards astronomical indicators – mostly Sun's annual movement on the horizon, which indicated seasonal changes, and which seemed to the minds of ancient peoples "simultaneously the cause and symbol of the world's great forces."[35]

The Triple Spiral, dated at 2400 B.C.E., inscribed on the inner chamber wall at 'Newgrange,' is lit up by direct sunlight at Winter Solstice, the annual seasonal point that traditionally celebrates the birth of light and form. As recently as 1969 C.E., Michael J. O'Kelly became the first archaeologist to directly observe and confirm that this event occurred.[36] At that time, not only was the event not understood but the possibility of it being deliberately designed to occur, was stubbornly resisted.[37] This new knowledge was not in keeping with previously held theories of these places being tombs, and having no relationship to seasonal movements of light and the telling of time: there was prejudice

---

33 Ibid., 18.

34 Michael Dames, *Ireland: A Sacred Journey* (Element Books, 2000), 192.

35 Brennan, *The Stones of Time*, 39.

36 Ibid., 36.

37 Paul Devereux, *Earth Memory: The Holistic Earth Mysteries Approach to Decoding Ancient Sacred Sites* (London: Quantum, 1991), 120.

that the Irish Indigenous mind could not have conceived such a cosmic picture.

It is significant that places like Newgrange, Stonehenge and Silbury Hill can only be comprehended when one is actually there and observes the relationship between the place and the cosmic/seasonal events – the Moon and Sun over a period of time, over years … and this is what the researchers had to do. These sites are living texts – that still speak when the receptive observant mind is present.

The similar lack of comprehension – by minds that could not get outside of their cultural frame – that was brought to bear on the monument of Silbury Hill, is documented by Michael Dames in *The Silbury Treasure.* The subtitle of that book is *The Great Goddess Rediscovered* – and that is the mindset that is required for the comprehension of this prehistoric structure, and also for the comprehension of the Triple Spiral. As Dr. Claire French points out most mythographers (and also archaeologists) have omitted "the importance and functions of the female deity and her possible role among the people of the Celtic realms."[38] The mind required for the real comprehension of art and monument from this period and place is one that allows the female metaphor for the sacred … Goddess, functions beyond the purely ornamental or an orgiastic "fertility cult."[39] And that mind has been hard to come by until recently.

The work of Marija Gimbutas has been notable and extensive in this regard. Gimbutas in her book *The Language of the Goddess*, notes the extensive use – all across Old Europe – of the "tri-line," appearing as early as 24,000 B.C.E.[40] She notes its later association with energy symbols like the whirl, or sometimes contained within uteri or seeds, or alternating with crescents, sometimes coming from the mouth or eyes of a Goddess, or with continuous *v* motifs on pottery: all indicating symbols of "beginning" and "becoming."[41]

---

[38] Claire French, *The Celtic Goddess: Great Queen or Demon Witch* (Edinburgh: Floris Books, 2001), 22.
[39] Ibid.
[40] Marija Gimbutas, *The Language of the Goddess* (New York: HarperCollins, 1991), 89.
[41] Ibid., 92-94.

Gimbutas notes the link of this motif with the Triple Goddess, "an astonishingly long-lived image" documented as early as 15,000 B.C.E. … "continuous throughout pre-history and history" down to the Greek and Roman, Irish, Germanic, Baltic and Slavic triple matrons.[42] Gimbutas says the repetition of the threes – in engravings and structure – at Newgrange Ireland is striking and seems to represent the Goddess as "the triple source of life energy necessary for the renewal of life,"[43] as Michael Dames later re-iterates.

Some researchers are now realizing that this Triple Spiral was inscribed by minds that understood their Place – Earth and Cosmos – as Mother-Creator: this place itself as the Sacred Entity.[44] For these ancestors of Indigenous Western tradition – Pagans – the Land 'itself'/ (Herself) was Mother Deity. And in our times, as the sense of this Land/country is extended to a huge Universe, the Pagan person perhaps becomes 'PaGaian': as we come to know Gaia in Her whole Cosmic and evolutionary context. And for these ancient forebears, She was triple-faced – there was the aspect of the whole – Ériu … Mother; the aspect of the differentiated individual parts – Fódla … Virgin; and the aspect of the hidden sentience within – Banba … Crone.[45] In the representation as tri-line or as Triple Spiral, She is not anthropomorphised – but is expressed and apparently understood as a cosmic dynamic – one that is suggested to be creative in an essential way.

Humans have for a long time identified a Creative Triplicity that runs through every part of the Universe. Author Caitlin Matthews identifies it amongst the Celtic peoples and says that this "innate triplicity" known as the Triskele, "may grace our lives with an ever-living energy."[46] It is also embodied in every breath – as it waxes, peaks and

---

[42] Ibid., 97.

[43] Ibid.

[44] I use capitals to designate these terms as names for Deity.

[45] As Michael Dames describes the aspects, *Ireland*, 192, with my addition of associated Triple Goddess phases. The naming of these phases as "Virgin, Mother, Crone" is only one possible way to name them, and generally does require a re-storying of each, before they may be understood with full integrity: see Chapter 1 of this book.

[46] Matthews, *The Celtic Spirit*, 366.

wanes – as all breath-taking beings do. This ubiquitous Triple-faced Creative Dynamic is most likely lunar in its original representation – as humans witnessed the Moon waxing into fullness, waning into darkness, and re-emerging.[47]

Martin Brennan, who spent long periods of time in the context and place of Brú na Bóinne observing and documenting events there, says that the Triple Spiral is "perhaps the most powerful representation" of the sacred heritage of ceremonial celebration of eternal creation represented in the Wheel of the Year, the phases of the Moon and the lives of all beings.[48]

The ceremonial celebration of the Seasonal Wheel of the Year may be an embodiment and year-long celebration of the Creativity that unfolds the Cosmos: a celebration of Cosmogenesis – the Creativity in which we live everyday. Cosmogenesis, in brief, is the ongoing creative activity of the Cosmos, the unfolding/evolution of the Universe – referring to the form producing dynamics of the Cosmos.[49]

In recent work, cultural historian Thomas Berry and mathematical cosmologist Brian Swimme, state that Cosmogenesis will be characterised by three governing themes, that "the Universe arises into being as spontaneities governed" by these three "primordial orderings," that the "very existence of the universe rests on the power" of these orderings:[50] which are by their awesome nature, beyond any one-line univocal definition. In their discussion of the three features, Swimme and Berry cover the full gamut of creative manifestation – from particles to biological life to stars. They name the three faces of Cosmogenesis as differentiation, communion, and autopoiesis: offering many synonyms for these terms: and also leaving room for the deepening and altering of these three "as future experience expands our present understandings."[51]

---

[47] Anne Baring and Jules Cashford, *The Myth of the Goddess: Evolution of an Image.* (Penguin Group, 1993), 18 and 596.

[48] http://gofree.indigo.ie/~thall/newgrange.html.

[49] Swimme and Berry, *The Universe Story*, 70.

[50] Ibid., 72.

[51] Ibid.

The terms may be summarized as follows:

differentiation – to be is to be unique
communion – to be is to be related
autopoiesis – to be is to be a centre of creativity[52]

There is nothing simple about the defining of these characteristics nor about the defining of waxing, peaking and waning, creation-preservation-destruction – and one cannot Be without the others. They are best imagined with a Poetic mind or "fuzzy" logic,[53] for they are deeply complex, interactive and multivalent. I have augmented my understanding of the Triple-faced Creative Dynamic with Thomas Berry's three features of Cosmogenesis[54] in this way:

Differentiation – to be is to be unique ... it is the Universe's "outrageous bias for the novel,"[55] associated with the Virgin/Young One aspect. It is the differentiated parts, the Urge to Be, felt in the beginning of the breath. This is the budding of new leaves, shoots of early Spring, the first sliver of the new Moon, your tentative new beginnings – at any time in your life.

Communion – to be is to be related ... "even before a first interaction ... the primal togetherness of things,"[56] associated with the Mother aspect. It is the whole, the Place of Being – felt in the peaking of the breath – the dynamic reciprocal interchange. This is the fullness of the flower, the ripeness of the fruit, the fullness of the tide and Moon, your

---

52 This summary definition is on the editorial page of issues of *Original Blessing*, 1990's, Matthew Fox.
53 See Vladimir Dimitrov and Bob Hodge, "Dynamic Character of Fuzziness" in *Social Fuzziology: Study of Fuzziness of Social Complexity* (Heidelberg New York: Physica-Verlag, 2002) and Vladimir Dimitrov, *Introduction to Fuzziology: Study of Fuzziness of Knowing* (Lulu.com, 2005).
54 See Appendix A, principle 4. Also see Thomas Berry, *The Dream of the Earth* (San Francisco: Sierra Club Books, 1990), 44-46, where he originally named "autopoiesis" as "subjectivity".
55 Swimme and Berry, *The Universe Story*, 74.
56 Ibid., 78.

creative engagement peaking.

Autopoiesis – to be is to be a centre of creativity … "the interior dimension of things … the power each thing has to participate directly in the cosmos-creating endeavour,"[57] associated with the Crone/Old One aspect. It is the hidden sentience within, the transformer. It is the creating of Space to Be – felt in the release of the breath. This is the seedpod forming, the grinding of the grain, the Autumn decline, the waxing Dark of the Moon, your letting go of your achievements, your small self dissolving into Larger Self.

The story – the unfolding – of the Universe, of Gaia/Earth, of Creativity, is an interaction of these three dynamics. To be attuned to these dynamics, as one may be in ritual/ceremonial celebration of them, is to be attuned to the everyday Creativity of the Cosmos, wherein conscious deep transformation is possible – the work of the 'Crone' aspect, conscious deep relationship is possible – the work of the 'Mother' aspect, the astounding beauty of particular self is known – the work of the 'Virgin' aspect. And one doesn't have to go anywhere – it is as close as your breath.

We may develop ethics and action based on these principles – we may "develop community, support diversity, and treat each other as sacred subjects."[58] The Triple Spiral may represent a powerful indigenous Earth-based jurisprudence – law – that is not separate from its representation of Being itself. Thomas Berry described three rights of "every component of Earth community, both living and non-living": "the right to be, the right to habitat or a place to be, and the right to fulfil its role in the ever-renewing processes of the Earth community."[59]

The Seasonal Wheel of the Year may be celebrated as an embodiment of this triple dynamic – like Gaia's Breath … waxing, peaking and waning, and re-emerging. The ceremonial celebration of the annual cycle of Earth-Sun Creativity – our everyday Sacred Site – may be

---

57 Ibid., 75.

58 Sarah Pirtle, "A Cosmology of Peace," EarthLight, Volume 15, No. 1, Fall 2005: 10.

59 Thomas Berry, *Evening Thoughts: Reflecting on Earth as Sacred Community*, ed. Mary Evelyn Tucker (San Francisco: Sierra Club Books, 2006), 149.

an expression of the dance of form and dissolution, that eternal dance in which we participate.

Adam McLean, primarily a researcher of alchemy, who spent much time studying and contemplating the Triple Goddess , describes relationship/alignment with this Dynamic as releasing "such a powerful current of creative energy as few have ever experienced."[60] He contends that She "remains a key to unlocking the store of ancient energies and spiritual wisdom" bound up within.[61] He speaks of the complexity of her guises, how She challenges our usual thinking with seeming contradictions and inconsistencies, yet he senses that She holds within Her all polarities – an integration of being that seems necessary for the "spiritual energies of the future" as he describes.[62]

Celebrating the Triple Dynamic in the Wheel

Over decades my work has been to enter into these three faces/qualities of the Sacred, to unfold an understanding of them. I have done this largely through ceremonial celebration of the eight Seasonal Moments: the process itself has in-formed me, taught me. (The ceremonial celebrations of Earth-Sun Creativity as I have scripted them, are based in the religious practice of Old Western European indigenous tradition, wherein there may be eight annual Earth holy days, though not all eight have always been marked in varying places and times.) The eight holy days ('holidays') are traditionally known as "Sabbats." I name them Seasonal Moments, taking inspiration from Thomas Berry's naming of seasonal and diurnal transitions as "moments of grace."

(Each of the Seasonal Moments as I have celebrated them, is based in theme, on the traditional understanding of it as articulated by Starhawk in *The Spiral Dance: A Rebirth of the Ancient Religion of the Great Goddess* (1979): it is a pre-Celtic cosmology.) I have adapted the Wheel of this tradition to the Southern Hemispheric seasonal dates, as well as in the way I tell the story. I have adapted the celebrations significantly, re-languaging them for myself and for different groups of people, and as

---

60 McLean, *The Triple Goddess* (Grand Rapids MI: Phanes Press, 1989), 12.

61 Ibid., 17.

62 Ibid., 12.

my understanding deepens and changes.

Different names may be given to the Seasonal Moments. I mostly use the Celtic names, though it's not necessary. There are layers of story associated with each point/season depending on time and place. Various tribes/regions have given different names to the points, and storied them differently … depending on whether agricultural or pastoral, and the extent to which the God overtook Goddess as metaphor. Sometimes the story is quite imbued with war metaphor – revealing a later period of human history. The story that is told reveals place, time and economy.

In the oldest stories in Indigenous Western tradition, the Earth cycle was a celebration of the Great Goddess Mother of All and of Her creativity frequently expressed in three phases of waxing, peaking and waning, in never-ending renewal: for this assertion I refer to the work of Marija Gimbutas (1982, 1991), Michael Dames (1976), Adam Mclean (1979, 1989), Charlene Spretnak (1992), Susan Gray (1999), Baring and Cashford (1993), Joseph Campbell, Caitlin and John Matthews (1994, 2000).

I story it in a manner kin to the Old way – as a dynamic of change in the three faces of the Creative Dynamic: it is a pattern of wholeness not a duality. The year long pattern becomes a celebration of Self, Other and Cosmos in complex relationship – no matter what your gender/sex or species.

In later stories of this seasonal cycle the God came to play more of a role in the seasonal story, and most often dis-placing the Goddess images as the dominant story – even in Pagan tradition. Then Christian metaphor has been added in more recent times of the last couple of millennia: sometimes substantially altering the sense of the Moment. In many places there is a complex mix of story and symbol from all the different layers – and people aren't conscious of the various influences on their cosmology. Often the earlier ones, that use female metaphor, are not mentioned or remembered – and that has been my work and passion … the recovery, speaking and celebrating of Her.

Each of these seasonal points has layers of story – the most ancient ones (still being uncovered), the Celtic ones, the Christian ones, and now in our time – our own new cosmological understandings. I tell you all that because you have probably heard many different stories, and also, so that you may get a sense of how the story can be made your own

… in this time and space. It may be a point of your relationship with Gaia-Goddess-Cosmos, in THIS Place and Time, and your particular life and community. It is a pathway into the Centre – an indigenous spiritual/religious practice.

Not only do names and story vary for different ethnic groups over the centuries, the actual dates are moved around a bit depending on the preferences of the particular group and perceptions and needs. The ceremonial dates were not originally based on the Gregorian calendar, as many assume today: that's what the stones and their alignments with the horizon, were for … the telling of the time, the approach of the Moment.

In summary, as has been celebrated in PaGaian sacred ceremony for decades: the light part of the cycle is about coming into being, celebrating what is commonly called the Virgin phase – which is understood as new differentiated being. Early Spring and High Spring ceremony in this light part of the cycle are about the beauty and joy of particular manifest form and its process.

The dark part of the cycle is about returning to the Great Plenum whence all emerges, celebrating what is commonly called the Crone phase – which is about interiority and deep creativity. Early Autumn and Deep Autumn ceremony in this dark part of the cycle are about the beauty and grief of transformation and its process – that can be more of a challenge: we are often filled with hubris and take it all so personally.

The Solstices, which are about the fullness of dark or light, and also the interchange from one to the other, may celebrate the Mother phase in particular – the relationship that this Place is – between the 'manifest' reality (of light) and the 'manifesting' reality (of dark).[63]

The Equinoxes are points of balance, in both the dark and in the light phase, wherein we may pause a moment and recognize the presence and power of all Three – the Sacred Balance. In PaGaian ceremony I have chosen the ancient story of Demeter (as Mother) and Persephone (who embodies both Young and Old One) – the Mysteries of Eleusis – to celebrate the Sacred balance of joy and grief, of being and loss, and the continuity of life. The Equinoxes, and the ancient icons of Demeter and Persephone, may express the delicate 'curvature of space-time,' the

---

[63] I take these terms from David Abram, *The Spell of the Sensuous*, 191.

fertile balance of tensions which enables it All.

Concepts Essential to PaGaian Cosmology
Most of the following are further developed throughout the following chapters.
(i) we are all native to this Place – "indigenous"

We are all native to this Place we inhabit – *Where we are* is *Who we are.* We have been turned into "outsiders" in our own land: most have been "colonized" in this way … perhaps especially women/beings-female and beings-dark, in the last few millennia: but others have been included in that colonization.

Other words instead of "indigenous" could be used – like endogenous (that is, all originate from within the organism of Earth), autochthonic (that is, all originate in this place of Earth and Cosmos where we are found): but the understanding is that all beings are Native; that is, each being's heritage in every cell goes all the way back to the first cell and then actually all the way back to Origins. Also one's "own land" is the place within one's own skin as much as anywhere, as well as the actual Earth, and the Cosmos: when I speak of *Place*, I mean *this* Place (bodymind) as much as anywhere. We have grown very used to locating Ultimate Reality outside of ourselves – instead of understanding that we are immersed in it. There is no "outside." This is true for us individually, but also ecologically and cosmically: these are all layers of the self – seamless – when we are able to perceive it in truth. This perception is perhaps the main aim of the process of ceremonial practice of PaGaian Cosmology: that is, for me, "indigeneity" means discovering that YOU/I/WE and each being is a Place of the Cosmological Unfolding.[64] The point is the identifying of one's own "Nativity," one's intimate embedded relationship with this Cosmos, this Earth. I personally have been finding my Place, *situating* myself with this process.[65]

And all have indigenous ancestors, though apparently many have forgotten: all have a lineage from particular regions of Earth, ancestors who knew their primordial connection to Land/Earth/Cosmos. Thus,

---

[64] This is expressed in the "Triple Goddess Breath Meditation": see Appendix J.
[65] As I have described in *PaGaian Cosmology*, 4-5.

all are grounded in some place. Some use the term "Pagan" to speak of this, particularly if from a Western European heritage. Andras Corben-Arthen has addressed this issue in some depth, and especially did so at the Parliament of the World's Religions 2009.[66]

(ii) an evolutionary perspective … the Universe Story.

By "evolutionary" I am referring to Cosmogenesis, the creative dynamic unfolding of the Universe. Sometimes "evolution" may be mistaken to imply "progress" of some kind. When the term "evolution" or "evolutionary" is used in PaGaian Cosmology, it is not meant to imply "progress" or any kind of "teleology" necessarily. It is simply meant in the sense of perceiving "a time-developmental process": that is, there is a story of the universe, a sequence of events that Western science has perceived as having taken place, and it is an "irreversible sequence of transformations," as Swimme and Berry describe.[67] There does appear to be greater complexity, greater variety and intensity as can be observed on planet Earth, but this does not mean to imply an "ascent" as has been common to think in Western culture. I do not mean "evolution" to imply a hierarchy of development, simply perhaps a holarchy – which is an expansive nested reality contingent on what went before.

(iii) we live in an Omnicentric Universe

This is the understanding that the Centre of the Cosmos is here as much as anywhere. In an omnicentric Universe, the Centre is everywhere. One very simple way of understanding this is with this story: "From this point, every direction is Infinity,"[68] but my understanding of it comes from Brian Swimme who describes omnicentricity in *The Hidden*

---

[66] This interview is an introduction to the discussion http://ecer-org.eu/interview-with-andras-corban-arthen-by-christopher-blackwell/.
[67] Swimme and Berry, *The Universe Story*, 223.
[68] From a sign at a BP station north of Bunbury, Western Australia. The full message was: "YOU ARE HERE @ THE CENTRE OF THE UNIVERSE. From this point every direction is INFINITY. When you leave this place, wherever you go on the Highway of Life, may you always *Be All There.*"

Heart of the Cosmos, under the title of "A Multiplicity of Centers."[69] To my mind it also implies that all share the same Origins, all are continuous with Origins. We are as buds are to the tree. If ever the thread had been severed, we wouldn't be here. Origins are ever-present. Omnicentricity is also implied in the Cosmogenetic Principle, which states that the dynamics of evolution are the same at every point in the Universe.

> What that means, amongst many other things, is that the same Creative principle that gives birth to the Universe, pervades every drop of it with the same creative potency – that the Centre of the Universe is everywhere. Thus it is here as much as anywhere.[70]

(iv) we live in a reciprocal Universe

The nature of the Universe, of the learning, and of the Journey is participative. We are in relationship with the perceived: we cannot touch without being touched at the same time. We are being watched – we are not (the) only observers. David Abram describes this so very well throughout his book *The Spell of the Sensuous.* The "giver" or "do-er" is always receiving. The receiving may be the gift. I feel that this quality of reciprocity is expressed in the emergent Goddess icon on the front cover of *PaGaian Cosmology* and on this book – She is being helped from the water, but/and She is emerging for their joy and delight. It is already a party![71]

My partner Robert (Taffy) Seaborne has summed up that there are three "R's" to reciprocity: receiving, relating, and returning. These also resonate with the three qualities of Cosmogenesis/Triple Goddess.

(v) What I mean by "Gaia"

The name of Gaia for Earth was popularised in our times by the scientific hypothesis that Earth was a whole living being – the "Gaia Hypothesis" put forward by Lynn Margulis and James Lovelock in 1974.

---

69 Brian Swimme, *The Hidden Heart of the Cosmos* (New York: Orbis, 1996), 80-89.

70 Livingstone, *PaGaian Cosmology*, 47-48.

71 When I chose this as the icon that represents my work, I was struck by the contrast of this to the image of a dead god on a cross.

It has been upgraded to a "theory," which means that "Gaian" research may proceed credibly within the scientific community. Before this popularisation there have been other scientists, poets and philosophers that have put forward this notion over the millennia and notably Vladimir Vernadsky in his book, *The Biosphere,* first published in Russia in 1926.[72] But the earliest ancestors were scientists: that is, they were keenly observant of the world around them. Thus, religion and science were in earlier times always at home with each other.

The concept of "Gaia" was not new for humanity but it was new for Western science – that Earth might be a whole system/being – possibly "alive"! Gaia is an ancient Greek name for Earth (as Deity) – with the understanding that She was alive. Some find the use of the term too Eurocentric: there are many names for the Primordial Mother from cultures around the globe. "Earth" itself is a Germanic Goddess name.[73]

Charlene Spretnak coined the term "Gaian spirituality" decades ago and meant it as "glimpsing the oneness of the sacred whole."[74] Our place of being is wholistic … there are no addendums or "away."

My understanding is that Earth-Gaia is not separate from Universe-Gaia. Earth is immersed in Universe. There is no seam that separates Earth-Gaia from Universe-Gaia. There are no heavens "up" there – we are IN it! Gaia can be known, felt, in any single articulation of Herself – within any Self. Gaia is a nested reality – Universe, Earth, Self.[75]

(vi) Concept of "holarchy"

A holarchy is a nested reality. Each layer of a holarchy – a

---

[72] Also referred to by evolutionary biologist Elisabet Sahtouris in *EarthDance: Living Systems in Evolution*. (Lincoln NE: iUniverse Press, 2000).

[73] See Glenys Livingstone, "Gaia: Dynamic, Diverse, Source and Place of Being" in *Goddesses in World Culture Vol 2*, Patricia Monaghan ed. (Praeger 2011), 143-154, also an edited version at https://pagaian.org/articles/gaia-dynamic-place-of-being/.

[74] Charlene Spretnak, "Gaian Spirituality." *Woman of Power,* Issue 20, Spring 1991, 10 -17: 17.

[75] Livingstone, *PaGaian Cosmology*, 29-36.

"holon" – has its own autonomy and is at once embedded within larger holons on which its existence depends … any small holon must synchronize its *autonomy* with the *holonomy* of the larger holon.[76] So celebrating Gaia is celebrating these nested realities all at once:

- self … biological, historical, cultural small self … this bodymind
- Earth … this planet, Jewel in the Womb of space.
- Cosmos … Earth is not separate from Her context …

This is our Habitat – this is all "Gaia." This is our Land – none of it separate.

(vii) Poetry
I name what I do as "Poetry," not "theology" nor even "thea-logy," as explained in the Preface and in *PaGaian Cosmology.*[77]

The Earth tradition in which PaGaian Cosmology has its origins, was an oral tradition, which relied on its poets. It may be noted that the first among the attributes praised in the Great Goddess Brigid was Her function of poet – along with physician and smith-artisan. Poetry was considered a critical discipline in which to engage – it was the way in which the culture was passed on, the transmission of the sacred stories, the cosmology of a people told, the bearing of a tradition.

(viii) "fuzziology" is a scientific term for work by Valdimir Dimitrov. I have found it useful for understanding the three qualities of Cosmogenesis and the Triple Goddess – how the Three are in each other, not linear.[78]

(ix) Light and Dark – as "manifest" and "manifesting"

I think of the light part of the cycle as acknowledging and celebrating the "manifest" reality, and the dark part of the cycle as acknowledging and celebrating the "manifesting" reality (or "unmanifest" as I used to name it). I take these terms from an analysis of the Hopi language wherein there is a discernible distinction between

---

[76] See Livingstone, *PaGaian Cosmology*, 33-34 for further references.
[77] Ibid., 42-45.
[78] See Vladimir Dimitrov, *Introduction to Fuzziology.* http://www.zulenet.com/vladimirdimitrov/pages/fuzzycomplex.html.

two basic modalities of existence.[79] David Abram summarizes the meaning of these terms with:

> The 'manifested,' … is that aspect of phenomena already evident to our senses, while the 'manifesting' is that which is not yet explicit, not yet present to the senses, but which is assumed to be psychologically gathering itself toward manifestation within the depths of all sensible phenomena.[80]

Knowing this connection – of the worlds – is indigenous mind, though the "gathering itself" as I understand it is more than "psychologically gathering itself": it is that and more, it is "physical" too in the sense of "physics." To practice the ceremonies of all eight Seasonal Moments – the whole Wheel, and to "sit out" within a wheel of stones that represent this, is to be held in that Present/Presence, of past and future as present – in this dynamic place of being, in which we are.

(x) Language – two aspects:

(a) the keen sense of the importance of what we speak; thus how we spell our reality. For discussion of "Goddess" and "archetype," see *PaGaian Cosmology.*[81]

(b) I am not speaking of a "Divine Feminine" or some *part* of Deity: I have no idea what "feminine" actually is:[82] thus I choose "Female Metaphor" since metaphor is all we really have to speak of the Absolute, and "female" may be a very appropriate sex for that metaphor given the regenerative nature of the Universe we dwell in. For discussion of the role of metaphor, a functional cosmology, and the Female Metaphor see

---

[79] Abram, *The Spell of the Sensuous*, 191.

[80] Ibid., 192.

[81] Livingstone, *PaGaian Cosmology*, 39-40.

[82] This is discussed in *PaGaian Cosmology*, Chapter 2. For a more recent version see: https://www.magoism.net/2013/07/essay-part-1-the-terms-feminine-and-masculine-by-glenys-livingstone-ph-d/and https://www.magoism.net/2013/08/essay-part-2-the-terms-feminine-and-masculine-by-glenys-livingstone/.

PaGaian Cosmology.[83]

(xi) Geotherapy

This a term used by Brian Swimme. The Seasonal Moments are points of personal relationship to Gaia – your story connected to Hers, a personal journey to wholeness through the Creative Metaphor. The practice of the seasonal ceremonies are points of conversation with Gaia/Ge and are therapeutic.

It is also a *cultural/communal* practice that may enable us to develop our atrophied sensitivities and subtle perceptions[84] … awaken us to who we are, where we are and enable more respons-ible action. It may be understood as "ecological psychology," a psychology that enables as it acknowledges, deep participation in our Habitat. This process of seasonal ceremony saved me, and still does.

Worthy of note in this context of Geotherapy, is the recent interest of neuroscience in the mind-shifting affects of ritual/ceremonial practices.[85]

---

[83] Livingstone, *PaGaian Cosmology*, 19-27.
[84] See Charlene Spretnak, *States of Grace: The Recovery of Meaning in the Postmodern Age* (San Francisco: HarperCollins, 1993).
[85] One example here: *The New Archaic: Neuroscience, Spiritual Practice and Healing*, co-presented by Anne Benvenuti Ph.D., Elizabeth Davenport Ph.D., and Glenys Livingstone Ph.D. at the Parliament of the World Religions, Melbourne, 2009, https://pagaian.org/articles/pwr-the-new-archaic/.

CHAPTER 1
RESTORING DEA AND INVOKING HER CREATIVE TRIPLICITY

Almost every ancient culture's creation myth begins with Her.[86] In the beginning was the Matrix, and the Matrix was all there was. "Before creation a presence existed ...(which)... pervaded itself with unending motherhood."[87] This Matrix was not "feminine," in any stereotypical way, which would limit Her to a certain mode of being. She was beyond all pairs of opposites. As the beginning and end of all things, She contained it all – she was yin and yang, right and left, dark and light, linear and cyclic, immanent and transcendent. There was not an

---

86 There are many references for this statement, and complete referencing can be found in Chapter 3 of my doctoral dissertation, *The Female Metaphor – Virgin, Mother, Crone – of the Dynamic Cosmological Unfolding: Her Embodiment in Seasonal Ritual as Catalyst for Personal and Cultural Change* (University of Western Sydney, 2002). The particular combined threads as I weave them are influenced by so many at this point in time; thus specific references may be arbitrary. Also, objectivity and subjectivity are hard to separate (true for any text though not usually admitted): some of the story as I have come to tell it has arisen organically over the years of my own reflection and then later found affirmation from published academic researchers. There *are* also "objective" sources for the storying that I do: that is, sources whose information is based in archaeological and mythological research and reflection, and is able to be checked. Those references have met academic requirements. The prominent influences and sources for the story of Her that I tell at this point in time have been Marija Gimbutas, Hallie Iglehart Austen, Merlin Stone, Mary Daly, Joseph Campbell, Erich Neumann, Anne Baring and Jules Cashford, Charlene Spretnak, Barbara Walker, Geoffrey Ashe, Marina Warner, Esther M. Harding, Lawrence Durdin-Robertson, Patricia Monaghan and Miriam Robbins Dexter.

87 Lao Tzu, *The Way of Life*, trans. Witter Bynner, (New York: Capricorn Books, 1962), 40.

either/or. She was not carved up into bits, apportioned a certain fragment of being – She was a totality. She bore within herself all of the polarities. Ancient Mesopotamian texts praise Ishtar of Babylon for her strong, exalted, perfect decrees as Lawgiver, and for her passionate, lifegiving sexuality, all in the one paragraph.[88] As Vajravarahi, Goddess has been known as Mistress of all knowledge, which included her physical being – quite a deal more expansive than more recent academic understandings of "Master of Arts." One of Ishtar's titles has been translated as "Great Whore," but this falls far short of the original understanding. As Merlin Stone has pointed out, the use of words like "prostitute" or "harlot" or "whore" as a translation for "qadishtu" negates the sanctity of this priestly role and reveals an ethnocentric subjectivity on the part of the writer;[89] (the term *Hierodule* is suggested as more accurate by Anne Baring and Jules Cashford).[90] The patriarchal bias in the minds of the writers disabled their comprehension of a holy woman who was sexual. The use of the word "Whore" to label One who embodied the Mystery of the Universe, has enabled patriarchal religions to denigrate the Female Metaphor[91] for deity – oftentimes out of ignorance, sometimes with conscious intent.

As Isis of Egypt, the Great Goddess was "Mother of the Universe." This did not mean that there was a Father of whom she was partner, as most human minds of our time assume. This title meant that she was the One from whom all becoming arose. It meant that she was the Creator. Many minds get caught up with a perceived need to affirm the male role in reproduction; however, there has never been the same affirmation in the West at least, of the female role in reproduction when the God has been Creator. To comprehend Mother as Creator does not omit the integrity of male being, it simply re-instates the integrity of the female and her Creative capacity.

---

[88] Hallie Iglehart Austen, *The Heart of the Goddess* (Berkeley: Wingbow Press, 1990), 130.
[89] Merlin Stone, *When God Was a Woman* (London: Harvest/HBJ, 1978), 157.
[90] Baring and Cashford, *The Myth of the Goddess*, 197.
[91] I capitalize this term to signify it as referring to a sacred and original entity.

As Mut of Egypt, She possibly preceded Isis. Mut is described as existing when there was nothing, the oldest deity, She who gives birth, but was Herself not born of any. Hers was an ancient name for "Mother," and as such was understood to hold the complete cycle that supported life – an original trinity – beginning, fullness, and ending. Mut's hieroglyphic sign was "a design of three cauldrons, representing the Triple Womb."[92] "Mother" was not a mere passive vessel, nor was she limited to birthing and feeding aspects that later cultures allowed her; "Mother" was a wholistic title incorporating the beginning and the end. She was "Om," the letter of creation and "Omega," the letter of destruction,[93] long before Jesus was said to have described himself this way.

As Neith, She was described as the "Great Cow who gave birth to Ra" – the Sun itself; Her parthenogenetic nature recognised. She was the primal abyss out of whom all being arose. In later times, Neith's story was greatly diminished and She was assigned a father god, as were many Great Goddesses around the globe – Brahma was installed as the father of Sarasvati, Chenrezig the father of Tara.

As Inanna of Sumer, She was "primary one" for three thousand five hundred years. Her story of descent and return, death and resurrection, is the oldest story humans have of this heroic journey, and it influences the later stories of redeemer/wisdom figures such as Persephone, Orpheus, and Jesus. Inanna was known as Queen of Heaven. In one image, Her power was expressed with a crown of horns on Her head, Her foot on a lion, wings and thunderbolts sprouting from Her shoulders.[94] Sumerian priestess Enheduanna of the second millennium B.C.E., and first known poet in Western cultural story, celebrated and wrote erotically of the sacred marriage[95] – that of Inanna and her lover Dumuzi. It is one of the oldest surviving written records of the Sacred Marriage myth cycle;[96] and although Her sexuality is

---

[92] Walker, *The Woman's Encyclopedia of Myths and Secrets*, 702.
[93] See Walker, *The Woman's Encyclopedia of Myths and Secrets*, 546.
[94] See Austen, *The Heart of the Goddess*, 74.
[95] Judy Chicago, *The Dinner Party* (Hammondsworth: Penguin, 1996), 31.
[96] See Starhawk. *Truth or Dare: Encounters with Power, Authority, and Mystery.* (San Francisco: Harper and Row, 1990), 40-47.

celebrated, Inanna's story never included pregnancy, as Starhawk notes.

In Greece, perhaps as early as the Paleolithic era, the Divine Female was known as Nyx, Black Mother Night, "the primordial foundation of all manifested forms," who laid the Egg of creation.[97] She was the full Emptiness, the empty Fullness. Aristophanes later sang of Her, "Black-winged Night ... laid a wind-born egg, and as the seasons rolled, Forth sprang Love, the longed-for, shining with wings of gold."[98] Her Darkness was understood as "a depth of love," not a source of evil as later humans named Her.

As Aphrodite, She was said to be older than time. Aphrodite as humans once knew Her, was no mere sex goddess and She was not only Greek; She was associated with the most ancient Dea Syria. Aphrodite was indistinguishable from the Fates and their power – perhaps more powerful. She was "multivalent," and had many names. This was characteristic of most Goddesses because the religion of the time was oral, and the stories of the diverse manifestations of the Ultimate Principle linked and were embellished upon as humans told them and travelled. Aphrodite was associated with the sea and dolphins, childbirth and the energy that opens seeds, sexuality and the longing that draws creatures together. The Love that She embodied as it was once understood was a Love deep down in things; it could be expressed as an "allurement" intrinsic to the nature of the Universe.[99] The Orphics sang of Her:

> For all things are from You
> Who unites the cosmos.
> You will the three-fold fates
> You bring forth all things
> Whatever is in the heavens

---

[97] George, *Mysteries of the Dark Moon* (San Francisco: HarperCollins, 1992), 115 -119.

[98] Ibid., 115, quoting Aristophanes in *The Birds.*

[99] This description of Aphrodite is the coalition of the work of Brian Swimme and Charlene Spretnak, as described by Spretnak in *Lost Goddesses of Early Greece: a Collection of Pre-Hellenic Myths* (Boston: Beacon Press, 1992), xvi.

And in the much fruitful earth
And in the deep sea.[100]

Surely She who represented such a power, could be said to represent a fundamental cosmic dynamic. Scientists in the last few centuries have spoken of a basic dynamism of attraction in the universe that is primal, using the word "gravity" to point to it, but it remains fundamentally mysterious.[101] And what difference Hymns of this kind to the Psalms, which have been understood to praise the Divine – surely One who unites the cosmos and brings forth all things deserves the dignity of ultimate divine praise.

In China, one of Her names was Shin-Mu, described as "Mother of Perfect Intelligence," who "miraculously" conceived Her first child,[102] and then gave birth to 33,333 creatures.[103] In later patriarchal stories She was said to give birth to all these creatures though Her arms and breasts – without a vagina (thus accommodating misogynist notions of "purity").

As Tara, She was known from India to Ireland as the primal Goddess. Praise and knowledge of Her has survived in Tibetan Buddhism through the millennia. In Tantric Buddhism She is understood to be at once both transcendent and immanent, at the centre of the cycle of birth and death, pressing "toward consciousness and knowledge, transformation and illumination."[104]

As Prajnaparamita in the Tibetan Buddhist tradition, the Female Metaphor is transcendent Wisdom: recognized as "…'Mother of all the

---

100 Referred to as "Orphic Hymn" in the 1994 calendar *Celebrating Women's Spirituality*, Crossing Press, Freedom California, week April 4 – 10.

101 Brian Swimme, *The Universe is a Dragon* (Santa Fe: Bear & Co., 1984), 43.

102 The attribution of "miraculously" is itself a later patriarchal frame for the parthenogenetic capacity of the Source and Mother of all being: Her conceiving and birthing was/is innate … "endogenous" is the word for having "an internal cause or origin."

103 Walker, *The Woman's Encyclopedia of Myths and Secrets*, 933.

104 See Erich Neumann, *The Great Mother* (Princeton: Princeton University Press, 1974), 333-334.

Buddhas' because Buddha activity arises out of, results from, and is born from Wisdom."[105] Her space is not a passive place, it is understood to be fertile and vibrant.

As Vajravarahi, She has been offered praise in the following way:

OM! Veneration to you, noble Vajravarahi!
OM! Veneration to you, noble and unconquered!
Mother of the three worlds! Mistress of Knowledge! ...
OM! Veneration to you, Vajravarahi! Great Yogini!
Mistress of Love! She who moves through the air![106]

Vajravarahi is a face of the Fire of the Cosmos, the Dancer, the Unseen Shaper.[107] She represents the everchanging flow of energy.[108] She has been imagined as holding a sword of insight and discernment, and a cup of blood – the blood representing the life force and potential for renewal as any Goddess' blood does. Vajravarahi is a sharp, compassionate Intelligence, pervading all.

As Mago of East Asia, She is the "First Mother" and "Originator of all species on Earth": … "Mago signifies the Female, another name for the Creatrix of the universe."[109] She is addressed by many names: Triad Deity (Samsin), Grandmother or Crone (Halmi), Auspicious Goddess (Seongo) and more … all these names rooted in traditional Korean/East Asian culture.[110]

As Kali Ma, in the Hindu tradition, She is addressed as Supreme and Primordial, alone remaining as "One ineffable and inconceivable ... without beginning, multiform by the power of Maya, ... the Beginning of

---

[105] Rita Gross, "The Feminine Principle in Tibetan Vajrayana Buddhism." *The Journal of Transpersonal Psychology*, Vol.16 No.2, (1984), 179-192), 186.
[106] Austen, *The Heart of the Goddess*, 124, citing a poem to Vajravarahi from a Tibetan Art Calendar 1987, Wisdom Publications, Boston.
[107] This is a title I have coined from Brian Swimme's name for Fire as unseen cosmic shaping power, in *The Universe is a Green Dragon*, 127-139.
[108] Austen, *The Heart of the Goddess*, 124, quoting Tsultrim Allione.
[109] Hwang, *The Mago Way*, 9.
[110] Hwang, *The Mago Way*, 10.

all, Creatrix, Protectress and Destructress."[111] The great mystic Ramakrishna of the 19th century, was overwhelmed by passion to realize Her and said he could not bear the separation any longer.[112] When She did reveal Herself to him, he experienced "a limitless, infinite shining ocean of consciousness or spirit" – he was "panting for breath."[113]

As Demeter of Greece, She is Mother of the grain, of wheat – "corn" as it was known, which was understood to reveal the Mystery of Being and was the core symbol of the Eleusinian Mysteries celebrated annually. The 'Vision into the Abyss of the Seed,' was a vision of the Vulva – the Mother of all Life.[114] Demeter is always in relationship with Her Daughter Persephone – they are a union of the new reborn within and of the old. Demeter as Mother gives the sheaf of wheat to Persephone as Daughter (a holy title), passing on the Knowledge; representing the continuity, the unbroken thread of life. Mother Goddess and Daughter, in this way reveal the Mystery of the seed in the fruit, the fruit in the seed, eternal Creativity. The grain is both the beginning and the end of the cycle, and thereby may represent knowledge of life and death – Divine Wisdom; and it is also food, thus embodying all three aspects of Goddess – creation/beginning, sustaining/preserving, and dissolution/de-structuring. The bread that wheat becomes, sustains the human, who also eventually gives itself away becoming food for the Universe: we are She. Persephone, like Demeter – the Grain, "becomes the goddess of the three worlds: the earth, the underworld, and the heavens."[115] They and their initiates are thus eternal.

In the Christian tradition, Mary of Nazareth came to embody Goddess, as many recount. This has been so mythologically, and in the hearts and minds of the people, regardless of the ambivalent official postures by the Church. Mary became known as Moon Goddess, Star of the Sea, Our Lady and many other titles that recall more ancient Goddess

---

[111] Walker, *The Woman's Encyclopedia of Myths and Secrets*, 489, citing Sir John Woodroffe (trans), *Mahanirvanatra*, 47-50.

[112] Walker, *The Woman's Encyclopedia of Myths and Secrets*, 493.

[113] Ibid., citing Colin Wilson, *The Outsider*, 254.

[114] Lawrence Durdin-Robertson, *The Year of the Goddess* (Wellingborough: Aquarian Press, 1990), 166-167.

[115] Neumann, *The Great Mother*, 319.

roots. Mary has been the one to whom the people turned, certain of Her love and mercy.

To the Sumerians the Divine was Queen Nana, to the Romans "Anna Perenna." She is Al-Uzza of Mecca, Artemis of Ephesus, Anatis of Egypt, Eurynome of Africa, Coatlique of the Aztecs, Kunapipi of Australia. She is Rhea, Tellus, Ceres, Hera. The Female Metaphor has been known in innumerable ways and by innumerable names as humans tried to express their perception of the Great Mystery. She encompassed All. She has been present throughout the millennia in the myths, rituals, religions and poetry of humanity. She has been loved and revered.

Before She appeared in human form, there were stones, trees, pools, fruits and animals that She either lived in or were identified with Her or parts of Her. For many peoples the stones and rocks were Her bones, the vegetation Her hair. Poppies and pomegranates and other such many-seeded flora identified Her fertility and abundance. Grain/food could represent Her. The earth itself was understood as Her belly, the mountains as places of refuge, caves providing shelter for the unborn and the dead. Primal peoples everywhere at some time understood Earth Herself as Divine One, Deity – Mother. They languaged this in different ways. The pre-Celtic indigenous Europeans named Her – the Land – as Lady Sovereignty.[116] In South-East Asia, where She has been known as Mago, Earth is Her Stronghold, the primordial home.[117] In Greece and in the West, She has been known as Gaia.

Central to understanding the Female Metaphor, is understanding the sacredness of vessels, pots, containers. These objects were understood as representations of Her. Pots, urns, pitchers "made possible the long-term storage of oils and grains; the transforming of raw food into cooked; … also sometimes used to store the bones and ashes of the dead."[118] The vessel was felt as an extension of the female body that shaped life, carried the unborn, and provided nourishment. Kettle, oven, cauldron have to do with warmth and transformation; bowl, chalice and goblet are vessels of nourishment and their openness is

---

[116] See French, *The Celtic Goddess.*

[117] See Hwang, *The Mago Way.*

[118] Adrienne Rich, *Of Woman Born* (New York: Bantam, 1977), 85.

suggestive of gift. The making and decorating of pottery was among the primordial functions of woman, often with taboos imposed on men to prevent them from going near. In later periods of human culture, in Eleusis, Rome and Peru and elsewhere the sacred vessels were supervised by the priestesses. The chalice was the holy Cup, felt as Her power to give life. Riane Eisler in *The Chalice and the Blade*, compares the chalice's power to give life with that of the blade, which is the power to take life, and develops how this was borne out culturally. In Christianity, woman was denied the right to handle the vessel as chalice – a ritual metaphor for the huge transition that had taken place in the human understanding; it was as if the female body no longer belonged to the female.

Water was a central Goddess abode, as it nourished and transformed, and also contained. She was identified with the water birds and ducks. As Bird Goddess She was the life-giving force, nurturing the world with moisture, giving rain, the divine food – the very milk of Her breasts. So, our ancestors frequently featured breasts set in rain torrents on the jars that they made.[119]

The tree as container and shelter, and also sometimes bearer of nourishment as in the fruit-bearing tree, was a central vegetative presence of Goddess. The figuring of such a tree in a negative context in later religious stories of humanity was not an arbitrary matter – this tree, particularly a fruit tree, was understood by the people of that time to be bearer of the Female Metaphor – Dea. The story was clearly a political statement, as many researchers now suggest.

Some animals were identified as particularly potent with Her; the deer with its fast-growing antlers speaking of Her regenerative power, the toad with its pubic shape, the bull with its crescent shaped horns, the butterfly that emerged from its dark transformative space, the bear that so powerfully protected the young, the pig with its fast growing body and soft fats. The pig's identification with Goddess, with the Old Religion of the Land, is evidenced by its later denigration, and to taboos on its

---

119 See for example Marija Gimbutas, *The Goddesses and Gods of Old Europe* (Berkeley and Los Angeles: University of California Press, 1982), 113-121.

flesh.[120] Similarly, animals with which women have been "insulted" – cow, duck, hen – are animals once sacred to the Female. The snake was especially significant as symbolic of immortality, vitality and rejuvenation because of its shedding skin. The snake's intimacy with the earth, its knowledge of the darkness of the earth's womb as well as the light of the upper world, made it a symbol of power and wisdom. It was a Mother-power and wisdom that the later patriarchs rejected, as evinced in their artwork and literature. The treatment of the snake's knowledge in the Genesis myth may be understood as a direct reference to the Old Religion. In Christian art, Mary as Goddess is often depicted standing on the snake crushing it.

As the humans developed symbols, one of the earliest representations of Goddess was the downward pointing triangle, the pubic triangle. This was a recognition of the Source of life, the Gateway. Sometimes Goddess was depicted displaying her breasts, belly, genitalia, or entire naked body as a form of divine epiphany. Today, Western science has come to understand that the Universe is still rushing away from its birthplace, still expanding. The Mystery is still birthing. The Gateway still pours Itself forth. All of manifestation is divine epiphany – Her ecstatic irrepressible expression. This ancient Goddess symbol has been renewed empirically.

Central to the spirituality and understanding of Great Goddess is the recurrent cycle of birth and death, the immortal/never-ending-renewal process of creation and destruction. It is a cycle seen most clearly in the moon, with its waxing, fullness and waning, which also corresponds to the body cycle of menstruation. The constant flux of things is manifest everywhere, in the seasons, in breathing, in eating. This is the nature of Goddess, Her manifestation, Her play. Anthropomorphized, this cycle is Virgin/Young One, Mother/Creator, and Crone/Old One. In Her most ancient and powerful depictions, Great Goddess embodies all three aspects – not just one; for example, Artemis is not only depicted as Virgin, in some images She clearly represents Mother and Crone too. These three qualities of the cycle of Goddess, belong together, and together they constitute a wholeness. In

---

120 Walker, *The Woman's Encyclopedia of Myths and Secrets,* 112, referring to Salomon Reinach, *Orpheus*, 19-20.

actuality they cannot be separated; one phase cannot "be" on its own; that is, a moon cannot always be full, the leaves cannot fall off the tree unless they grew there first, a new breath cannot be taken unless the old one is expired. The cycle has these aspects, but it is One. And so, Goddess of old was known, a union of three faces, complete and whole, yet ever in flux and dynamic. This triple aspect metaphor was later used to describe the triune nature of the patriarchal God, in both the East and the West, though in the Western teachings of the trinitarian Deity, its relationship to the cycle of life was most often more abstract.[121]

Ultimately the Female Metaphor – Dea – Goddess, is about the celebration of life, its eruption, its flux, its sustenance, with all that life demands and gives. She is an affirmation of the power symbolized by the chalice, the power to give life: initiate it, sustain it, pour it out. This is the power to Be, that all beings must have; not the power to Rule, that only a few might take. The popular Jungian understanding of the "feminine" is not sufficient to contain Her, shuffled off as She usually is to a portion of reality. And frequently that portion in the popular mind has consisted of passive receptive and 'user-friendly' qualities. These qualities are only part of the whole picture. As Virgin, Mother and Crone, She is eagle, bear, lioness, snake, as well as deer, gentle breeze, flower, rabbit.[122] She is not manifesting "masculinity" when she hunts for food, and neither is the human female when she operates in the world analytically or assertively. She is Herself.

I continue to use the terms "Virgin, Mother, and Crone" as three possible names for the qualities of the Triple Goddess whom many have loved in Her different forms throughout the ages. In my opinion, the re-storying of these terms is still a useful exercise – to expand the reduced notions that have evolved over millennia of androcentric thinking and culture. In the last few decades, I sat with many women in circle and we told stories of our lives within the frame of "virgin/young one,

---

[121] A notable exception is where Jesus was characterised as the Green God, and this image portrayed on churches. See William Anderson, *Green Man: The Archetype of our Oneness with the Earth* (Helhoughton FAKENHAM: COMPASS books, 1998).

[122] See Gimbutas, *The Language of the Goddess*, 316-317 for a description of the wholeness by which "Goddess" was understood.

mother/creator, crone/old one"; and found it to be a means of reconstituting a larger, deeper and freer sense of being, as we recognised ultimate and omnipresent Creative Cosmic qualities within us. I have also created new names for this Creative Cosmic Triplicity: "Urge to Be/She Who Will Be," "Place of Being/She Who Is," and "She Who Creates the Space to Be/She Who Returns All." As qualities/themes of Cosmogenesis, She is multivalent. She may be understood poetically.

The Virgin/Young One Re-Storied

I have associated this aspect/quality with the *Urge to Be*; as such She is concerned primarily with *love of self*, with the advent of Her unique and differentiated being.

The Virgin as she has been known in patriarchal times is a distortion of the original understanding of Her. She is originally primarily in relationship with herself, and she is not asexual. She is decidedly self-determined, remains her own property, whether or not she has sexual relationships. The term *virginity* signifies *autonomy*, and is a power to be "at cause," instead of "at effect"; it is only in later patriarchal stories, that a Goddess' autonomy was "concomitant with a loss of her sexuality," as in Athena's case.[123] The Goddess of old was always considered virginal; it was an ever-present quality of Hers. Even in some later stories, before the quality was completely diminished, She frequently "renewed" her virginity ritually.[124]

Esther Harding expressed that,

> the woman who is virgin, one-in-herself, does what she does – not because of any desire to please, not to be liked, or to be approved, even by herself; not because of any desire to gain power over another ... but because what she does is true.[125]

---

[123] Miriam Robbins Dexter, *Whence the Goddesses: A Source Book.* (New York: Teacher's College Press, 1990), 143.

[124] Jane Ellen Harrison, *Prolegomena to the Study of Greek Religion* (New York: Meridian Books, 1957), 311-312, and Dexter, *Whence the Goddesses*, 167-170.

[125] Esther M Harding, *Woman's Mysteries* (London: Rider & Company, 1955), 125.

The Virgin's purity is this: Her unswerving commitment to Her truth, Her true self. This self-serving purity is a deep commitment to being. Later patriarchal obsessions with unbroken hymens, turned the Virgin's essential "Yes" to life into a "No."[126] She became reduced in Christian times to a "closed gate,"[127] sometimes naive. In the Olympian pantheon the Virgin often came to be associated with harshness and indifference.

(It was because of the Virgin's association with the beginning of things, the emergence of life, that She came to be understood as passionately protecting the flame of Being – "the 'hearth,' which is also the original altar."[128] She loved all beings, desired their existence. She knew Creative Lust – Lust for Being. So Virgin Goddesses have guarded perpetual flames, representing this purity of purpose and passion. Diana, Great Virgin of Rome, is depicted with a flame. The priestesses of Celtic Goddess Brigid in Ireland tended a flame; now tended by nuns and Brigid has been re-configured as a saint.)

As Artemis in Greece, in Her Virgin aspect She was revered as midwife because of her single-minded drive to bring life into being.[129] The earliest stories of Artemis speak of a Goddess for whom "each creature – each plant, each wood, each river – is ... a Thou, not an it."[130] Women called upon Her in childbirth, and the labour-easing herbs used by midwives in Old Europe were called *Artemisia.* Artemis came to be known as One that protected and nurtured the young and vulnerable, the will to life, the spirit[131] – as much concerned with physical being as with the making of soul; there was no separation. As Virgin, Artemis was

---

126 Audre Lorde, "Uses of the Erotic," in *Weaving the Visions: New Patterns in Feminist Spirituality.* Judith Plaskow & Carol Christ, eds. (New York: HarperCollins, 1989), 208–213, influenced my understanding of the *yes* within ourselves.

127 Marina Warner, *Alone of All her Sex* (New York: Alfred Knopf, 1976), 73.

128 Neumann, *The Great Mother*, 284-285.

129 Spretnak, *Lost Goddesses of Early Greece*, 77-79.

130 Christine Downing, *The Goddess: Mythological Images of the Feminine.* New York: Crossroad, 1984), 167.

131 Merlin Stone, *Ancient Mirrors of Womanhood Vol 1* (Boston: Beacon Press, 1984), 381-386.

associated with untamed nature, the pre-domesticated, the pre-informed, the wild. She was the possibility of the open mind, the new and untried. She had no need to be afraid, because She was certain of taking care of Herself. Artemis was known as a Mighty Huntress, and in earliest human cultures this was not contradictory to deep relationship with the animals that were hunted. She was also known as Lady of the Beasts; the deer was often Her animal – an animal associated with birth and renewal, and the bear associated with rebirth/hibernation and fierce mothering. Artemis is often depicted as an archer. Her arrow that flies true and on centre, is just as surely the arrow of Self.

Athena has, in Western secular culture, commonly embodied the patriarchal version of Virgin – depicted as She has been in a suit of armour. In Athena's story as it evolved over time, can be seen a story of women throughout the ages.[132] Originally Athena has strong connection with the North African Goddess Neith, a primordial "Virgin Mother, the Holy Parthenos."[133] In Her oldest images and stories, Athena was associated with bird and snake,[134] and was the inventor of all arts.[135] Patriarchal myth accounts for Athena's existence by virtue of Zeus giving birth to Her from his head, after having swallowed her mother Metis when Metis was pregnant with Athena. Metis, Goddess of Wisdom Herself, cannibalized by Zeus, was said to counsel Zeus from within his belly;[136] She was in effect, the first woman behind every great man. Athena became the archetype of the patriarchal dutiful daughter, used to give authority to her father's edicts that included the denigration of Her own kind. In the Oresteia, the frequently performed Greek drama, Athena casts the deciding vote to acquit Orestes of the murder of his mother. The grounds for his acquittal is that the mother is not a parent,

---

[132] Mary Daly, *Gyn/Ecology: The Metaethics of Radical Feminism*. London: The Women's Press, 1979,13-14.

[133] Marguerite Rigoglioso, *The Cult of Divine Birth in Ancient Greece* (New York:Palgrave Macmillan, 2009), 52.

[134] Marija Gimbutas, *The Living Goddesses (edited and supplemented by Miriam Robbins Dexter)*. Berkeley and Los Angeles: University of California Press, 1999.

[135] See Spretnak, *Lost Goddesses of Early Greece*, 97-101.

[136] Daly, *Gyn/Ecology*, 13.

merely the nurse of the male seed. Athena then persuades the Furies, the "last remaining representatives of woman's old powers" to submit to the new patriarchal order.[137] Whereas, in the older stories, Athena was daughter of the Mother, indistinguishable from the Mother Herself. She was spiritual warrior – protecting the arts and wisdom, not a soldier. Her holy quest had been in the service of life, urging forward the creative spirit. It was Her vision, not armour, which was Her strength.

The Virgin aspect loves Herself, as She loves all, identified as She is with life itself. To despise self is to despise All. As Aphrodite, She "lifts Her robe to admire her own full buttocks;"[138] Inanna too, Great Goddess of the Sumerian people explicitly rejoices in Her own sexual beauty. Aphrodite, like her Sumerian Sister, is the Creative Force itself. In Aphrodite's case, She is identified with the oceans as Source of Life, and doves and waterbirds attend Her; the inseparability of the Mother and Virgin aspects is obvious here. In contrast to this perspective on Aphrodite is that of Jungian, Robert Johnson:[139] after affirming that all women contain "the Aphrodite nature," Johnson proclaims "her chief characteristics ... (as) ... vanity, conniving, lust, fertility, and tyranny when she is crossed."[140] Charlene Spretnak wonders particularly about his inclusion of "fertility" in the "string of negative adjectives."

Persephone is a Virgin Goddess who has been to hell and back. In the earliest story of Persephone,[141] the Crone and the Virgin aspects are inseparable: Persephone chooses to go to the underworld and indeed becomes Sovereign. She comes to know this realm, to guide others through it, and is equally associated with re-emergence, re-generation. She is not a naive Virgin; she goes into the darkness in trust, knowing its fertility, and her own impetus to sprout afresh, to begin again. She has been around the block many times, and because of that, continues to believe in her capacity to take form again. This knowledge of the cycle

---

137 Riane Eisler, *The Chalice and the Blade* (San Francisco: Harper and Row,1987), 81.
138 Austen, *The Heart of the Goddess*, 132.
139 Robert Johnson, *She: Understanding Feminine Psychology* (New York: Harper and Row, 1977).
140 Spretnak, *Lost Goddesses of Early Greece*, 35 quoting Johnson, *She*, 6.
141 See Spretnak, *Lost Goddesses of Early Greece*, 105-118.

of life and death is the Mystery that was celebrated in the rites of Demeter and Persephone in Eleusis and in the earlier Thesmophoria.[142] It was so, long before the Paschal Mysteries of Jesus crept in. Persephone's descent is a return to the depths for Wisdom, and Her emergence from the Earth is witness to the power to be, that surges through all Creation continuously; and inseparably in individual beings. She is the Seed of Life that never fades away, an energy present in all: as such She tends the sorrows.

The anthropomorphic forms of Virgin named as Artemis, Athena, Brigid, Aphrodite, Mary and many more – so that we may speak of Her, signify an energy, a creative dynamic, an aspect of divine essence – whatever it is in the dead looking branch that pushes forth the green shoot. She can be *felt* as the Urge to take a new breath, as hunger for food, as hunger for anything. She is passionate. She can be felt in any longing. She midwifes the soul, and any creative project. She is known when there is self-love, one's beauty recognized, one's truth held firm. She is the hope, the Promise[143] of fulfillment – expressed in the image of the new crescent moon, and felt, as that fine sliver of light enters the eyes. She is all possibility within the bodymind, within the seething quantum foam.

I associate the Virgin with the Buddha nature, the Shining One within all,[144] that calls us forth … She is the future for whom we "refine the gold." Virgin nature is *"She Who will Be,"* who holds forth her song despite forces of disintegration. She is the courage, confidence and exuberance to say "yes" to each particular small self.

The Mother/Creator Re-Storied

I have associated this aspect/quality with the *Place of Being*, the web of life; as such She is concerned primarily with *love of other*, with the essential relationship and communion that being is.

---

[142] See Baring and Cashford, *The Myth of the Goddess*, 374-390.
[143] Starhawk's term for the Child/Young One, *The Spiral Dance*, 257.
[144] I acknowledge Joan Halifax, *Being With Dying,* CD series (Colorado: Sounds True, 1997), for a broadened understanding of the Buddha, the Sangha and the Dharma – which I now associate with the three faces of the Female Metaphor.

Where the Virgin/Young One is primarily in relationship with *self*, the Mother is primarily in relationship with *other*. She is the network of relatedness, the Weaver of the Fabric, the peaking of Creative power. As Mother, Goddess is primal – the first concept of divinity, the Creator. She is the beginning and end of all things, the Creative Force of the Universe, parthenogenetically giving birth to all life. Earliest humans had no reason or inclination to correlate meiotic sex with childbirth: women must have eventually noted it, becoming the keepers of a lunar based calendar, which coincided with menstrual cycles. The human community, the village, gathered around the primary dyad of mother and child. The woman as mother was perhaps the original civilizing force, though the current texts of most cultures record "civilization" as a male accomplishment, that motherhood distracted women from this. Motherhood most likely gave the very impetus to grow, sustain, beautify, count, write.[145]

She "wielded the digging stick or the hoe … tended the garden crops and … turned raw wild species into prolific and richly nutritious domestic varieties."[146] Indeed it is still she who comprises some eighty percent of the world's farmers. In Egyptian hieroglyphics, "house" or "town" may stand as symbols for "mother." Ancient civilizations traced their descent through the mother; the mother was the basis for the clan, often named after her. Some ancient tomb inscriptions disregarded fathers.[147] Even the earliest patriarchies remained aware of and prepared to admit the source of their power; it was only later that the mother became a merely passive vessel, particularly in Greek philosophy. Pharaohs ruled by matrilineal succession. The literal throne itself is a stylization of the lap of the Mother; this was the seat of power – to be taken on Her lap.[148] As menstruant and potentially mother, the female was the first measurer of time,[149] and observer of numerical relationships

---

145 Walker, *The Woman's Encyclopedia of Myths and Secrets,* 684-685.

146 Ibid. 681 quoting Wolfgang Lederer, *The Fear of Women* (NY: Harcourt Brace Jovanovich Inc., 1968), 87.

147 See Walker, *The Woman's Encyclopedia of Myths and Secrets,* 681-682.

148 Neumann, *The Great Mother,* 98-100.

149 Judy Grahn, *Blood, Bread, and Roses: How Menstruation Created the World* (Boston: Beacon Press 1993), 155-157.

and pattern – original mathematical skills.[150] The Sanskrit word "matra" and the Greek "meter" both mean "mother" and "measurement."[151]

Barbara Walker notes how religions based on the Mother were free of the quest for indefinable meaning in life, as such religions "never assumed that life would be required to justify itself."[152] Such religions, were generally free of guilt, fear and a sense of sin; since birth, not baptism was the only pre-requisite for belonging. Even patriarchal religions have reached for maternal imagery to describe the love of the God. Buddha too described Universal Love in this way:

> As a mother … protects and loves her child … so let a person cultivate love without measure toward the whole world, above, below, and around, unstinted, unmixed with any feeling of differing or opposing interests ... This state of mind is the best in the world.[153]

This maternal energy, seen here as the deep spiritual calling of all humans, has been for women in many cultures a zone of entrapment. Patriarchal religions have exhorted her to embody this unconditional love of the Other, with no balancing factor of love of Self. Where her capacity for this love of Other has been given its due, it has not at the same time, been recognized as a capacity for spiritual leadership. In the Catholic tradition, Mary is praised as a paradigm of virtue, and yet women and girls have sacramental roles withheld from them. If it is as perfect servant that she is praised, by their own theology she is perfect model for leadership. Mary is apparently the exceptional woman, yet Jesus is obviously the exceptional male – a fact that does not prevent male leadership. In the early days of Christianity, Mariology was a rival religion, with grassroots allegiance to Mary as Deity – so with the proclaiming of Mary as Mother of God in 431 at Ephesus, the church incorporated this

---

150 See Grahn, *Blood, Bread, and Roses* 157-171.

151 Walker, *The Woman's Encyclopedia of Myths and Secrets,* 685.

152 Ibid., 693.

153 Ibid., 694 referring to Ross, *Three Ways of Asian Wisdom* (New York: Simon & Schuster, 1966), 123.

powerful image[154] – rather like Zeus swallowing Goddess Metis, thus appropriating her wisdom and creative power, and ensuring the people's allegiance. Simone de Beauvoir said that "it was as mother that woman was fearsome; it is in maternity that she must be transfigured and enslaved."[155]

The fathers of the Christian church have been profoundly ambivalent about details of Mary's motherhood, and her relationship to divinity, Jesus and humanity; motherhood is a profoundly ambivalent role in many cultures. It is clear that the maternal energy is indeed something that would befit all humans to aspire to. It is a holy passion, but it has been appropriated; frequently woman is locked into domesticity or pouring life's energy into mothering an institution or a man from whom she is getting no return. The political and global significance of her consuming passion to sustain the world and make it better, has not been recognized,[156] and it has not been balanced with a consuming passion for her own being. The Mother's relationship to Other, Her Creative Power to give life, in earliest mythologies, was not the prison that She was later contained in.

In patriarchal mythologies throughout the world, the Mother of All has frequently become "wife" of some god. Hera is such a One: known in this recent era, as jealous, quarrelsome wife of Zeus, She pre-dated him by far as ancient face of the parthenogenetic Mother. The first "Olympic" races held every four years, had been Hers, with runners – all girls – selected from three age groups representing the lunar trinity.[157] Hera and Zeus' constant mythological quarrels reflected real conflicts between the early matristic cultures and the rising patriarchate. She and the Amazon queens who represented Her did not go quietly, and they

---

154 See Geoffrey Ashe, *The Virgin* (New York: Arkana, 1988).
155 Simone de Beauvoir, *The Second Sex*, translated by H.M. Parshley (New York: Knopf, 1953), 171.
156 This is evident in derogatory references frequently made in political discussions, to "motherhood" statements and policies, as if "motherhood" describes the statement's/policy's small-mindedness and trivial nature. The word "trivial" itself has its roots in the sacredness of the "tri-via" – three way path.
157 Spretnak, *Lost Goddesses of Early Greece*, 88.

remained discontented with the new regimes. Hera's troublesome nature in the Olympic pantheon reflects One who had been "coerced but never really subdued by an alien conqueror."[158]

In some indigenous traditions around the globe, the birthing mother is understood as a model for courage – Native American, Samoan and Aztec cultures honoured her as warrior.[159] Many ancient cultures understood birthing as a ritual act; in Catal Huyuk there is a ceremonial birthing shrine "with red-painted floors and images of the ubiquitous Open-legged Goddess in labor."[160] Vicki Noble describes the birthing woman as "quintessentially shamanic,"[161] for in this act, she goes to the gates of life and death and with the most intense encounter with universal forces, experiencing trance states, she brings another being into the world. Perhaps the phenomenon of "post-natal depression" in these times is a symptom of the lack of recognition of this.

Goddess as Mother is the Weaver of the Fabric of the Universe, with many ancient Goddesses imaged this way. This power came to be feared, rather than revered – in Her "character as creator, sustainer and increaser of life" the Great Goddess came to be seen as "negative and evil," by a consciousness that desired "permanence and not change, eternity and not transformation, law and not creative spontaneity …(turning) her into a demon."[162] This consciousness, which Neumann calls an "antivital fanaticism," feared being "ensnared" in the "web of life, the veil of Maya."[163] Sometimes the weaving activity of women became known as the cause of illness or a curse with some Christian traditions even forbidding knitting. Ixchel the Weaver, Mother, Queen, Grandmother to the peoples of Southern Mexico, the Yucatan Peninsula,

---

158 Ibid. 89 quoting Jane Ellen Harrison, *Themis: A Study of the Social Origins of Greek Religion* (Cambridge: Cambridge University Press, 1912), 491.

159 Austen, *The Heart of the Goddess*, 18.

160 Ibid, 20.

161 Vicki Noble in "Female Blood Roots of Shamanism" quoted by Austen, *The Heart of the Goddess*, 18.

162 Neumann, *The Great Mother,* 233.

163 Ibid.

and most of Central America,[164] came to be symbolized by an "overturned vessel of doom"[165] – yet "for centuries, women have made pilgrimages to Her holy places."[166] Austen describes Ixchel as sitting "at Her loom with Her ever-present bird companion, the nest weaver, who is associated with Goddess throughout the world:" Ixchel "easily and with great presence ... in the bliss of creativity" spins "from her deepest being" and breaths the "breath of eternity, ... the life force into each being."[167]

This Mother energy has been named "ten thousand names," as humans experienced and witnessed Her. She can be ***felt*** in the fullness of a breath, that dynamic interchange. Her power can be felt in the satisfaction of successful completion, in the successful tending of needs, in the holding of another being; when life is given in some form to another. The Mother can be felt in the comfort offered by needed rain, seen and smelt in the full flower, tasted in the ripe fruit. She is in the work of everyday, of strengthening networks, weaving and repairing, creating the world, raising children, teaching adults. She is Creativity in its fullness – expressed in the image of the full moon, and felt, as that awesome disc of light enters your eyes. She is the realization of passion, the bliss of union.

I associate the Mother with the Sangha, that complex supportive community around the globe, the interdependent web without whom none would be sustained.[168] She is the present, the eternal now, the living of life as if it goes on forever – and indeed it does; if the thread were once broken, none would be here. The form changes – every atom recycled infinite times, the shape shifts, but life goes on. She is the pure gift of every moment, filled as each moment is with the Creativity of the whole web since the beginning. She is *"She Who Is."*

---

[164] Austen, *The Heart of the Goddess*, 10.
[165] Neumann, *The Great Mother,* 187.
[166] Austen, *The Heart of the Goddess*, 10.
[167] Ibid.
[168] I acknowledge Joan Halifax, *Being With Dying,* for a broadened understanding of the Buddha, the Sangha and the Dharma – which I now associate with the three faces of the Female Metaphor.

The Old One/Crone Re-Storied

I have associated this aspect/quality with *She Who Creates the Space to Be*; as such She is concerned primarily with *love of All-That-Is*, dissolving the old in every moment, ever transforming, and essentially creatively fertile.

The Old One as a phase of Being is significant. The fact that life continues past reproductive fertility for the human woman indicates an evolutionary interest in the creativity of this phase; in humans as in all.

The celebration of the threefold cycle of Goddess is about the celebration of life; and it includes the waning of being, death and the darkness, in its full picture. "The color black, now commonly associated with death or evil in Christian iconography, was in Old Europe the color of fertility and the soil."[169] Over the centuries, the Christian mind has imagined the religion of the Female/Goddess to be sordidly pre-occupied with death, whereas in fact the reverse is true: it is in the denial of decline and death that we become surrounded with it, because there is no place for new life to spring from. In a cultural context where people imagine or pretend that they are immortal, where death/darkness is seen as abnormal, it becomes a fascination, and creates a planet that is over-burdened with waste. When the end part of the cycle is given a place, its reality included, there is very little "garbage." Such a mind comprehends that it is in the compost, the de-composition, in the darkness, that new life is nurtured, fertility is found. It is in the acceptance of death that wisdom is gained, and life is lived more fully. When the Female Metaphor was whole, death was not understood as separate: artist Patricia Reis points out that for 30,000 years (33,000 B.C.E. – 3,000 B.C.E.) there were no images of a horrific Goddess.[170]

The Crone/Old One is primarily in relationship with All-That-Is (where Virgin/Young One is primarily with Self, and Mother/Creator is with Other). The Crone/Old One is that movement back into the Great Sentience out of which All arises, thus She sees into the elements behind form. She is often depicted with wide open eyes; often associated with the gaze of owl or snake – and knowledge of the Dark. In Her Egyptian

---

[169] Gimbutas, *The Language of the Goddess,* 144.

[170] Patricia Reis, "The Dark Goddess," *Woman of Power*, Issue 8, Winter 1988, 24-27: 24.

form of Maat, She was known as the "All-Seeing Eye"[171] and Maat's plume/feather was the hieroglyph for "truth."[172] This aspect of Goddess is the *Wisdom of the Ages* – and beyond all knowledge. The Crone aspect is the contraction that initiates destruction, when structure is no longer life-serving. Her contraction may also be understood as a systole, a contraction of the heart by which the blood is forced onward and the circulation enabled.[173] She is the systole that carries all away – She is about loss, but the contraction of the heart is obviously a creative one, it is the pulse of life. Is it only our short-sightedness that keeps us from seeing the contraction that way? Is it because we insist on taking it all so personally, when it is not, in fact?

As the end of the cycle, She is known in the breakdown of the endometrium, the shedding of the old, the flow of menstrual blood. She may be represented holding a bowl of blood, as Kali is, which signifies womb potency. The blood of Goddess is "'self-produced, primeval matter' the ocean of uterine blood before creation, holding future forms in the condition of formlessness or Chaos."[174] It is not blood that is shed by the blade, it is blood that naturally flows in the cycle of life: it is regenerative. In the patriarchal narrative menstruation has been a source of shame and pain for women,[175] and blood shed by violence has been much preferred to that shed by the female. Violent blood shedding became a form of entertainment, as well as a grasp for power. Its mythic advent can be traced to the Epic of Gilgamesh in the second millennium B.C.E.:[176] the hero grasps for a *synthetic* power – a substitute for the power to give life. The *organic* blood of the Female is indeed worthy of awe,

---

171 Walker, *The Woman's Encyclopedia of Myths and Secrets,* 294.
172 Ibid., 561.
173 I owe the "systole" metaphor to Loren Eiseley, *The Immense Journey* (New York: Vintage Books, 1957), 20. He uses it to speak of an eternal pulse that lifts Himalayas, and then carries them away.
174 Walker, *The Woman's Encyclopedia of Myths and Secrets,* 723.
175 See Penelope Shuttle and Peter Redgrove, *The Wise Wound: Menstruation and Everywoman* (London: Paladin Books, 1986) for a re-storying of menstruation. Also DenaTaylor, *Red Flower: Rethinking Menstruation* (Freedom CA: Crossing Press, 1988).
176 Starhawk, *Truth or Dare,* 47-60.

indicating as it does, a dynamic of loss in the nature of things, but the larger arc of that dynamic is consistently creative. Flesh, like all matter, is in constant flux. "Creation postulates change and any change destroys what went before."[177] The Crone is the one who "clears the decks," without which the new is not possible. She and the Virgin/Young One are always linked, the end and the beginning; One cannot exist without the Other. The snake that sheds its old skin is Goddess' symbol of constant decomposition, constant renewal. It is in the burning that the fire creates warmth and light. The lioness kills to feed her young. It is in the eating that teeth and bodies break down what is needed for sustenance – the Crone/Old One is our constant companion.

Because of the snake's association with power, wisdom, transformation and renewal it came to be associated particularly with the dark aspect of Goddess, but originally the snake/serpent *was* the Great Goddess herself: She was whole. In Jewish and Christian traditions of the West the snake came to symbolise evil. The Christian Goddess Mary, has been depicted as standing on the serpent, crushing it, in opposition to Her ancient heritage; yet it also presents a visual re-association.

Amongst names for the Crone quality are Hecate, Kali, Cailleach, Hel, Lilith, Medusa, Coatlique, Chamunda and Selket. These Goddesses are associated with death, devouring, seduction, rebellion, anger, darkness and awesome power – usually in a negative sense, though not always. Yet in the Christianized West, She almost completely disappeared; Her remnant is the "wicked witch/hag" of children's stories and cartoons, whose potency and intelligence is frequently belittled. "Old woman" became a term of derogation in the Western cultural context, meant to reflect uselessness. Miriam Robbins Dexter points out that although patriarchal cultures could find a place for the use of the virgin and mother energies, they could find no such use for the old woman.[178] The young virgin could represent "stored energy," and thus she maintained some numinosity; the mother "transmitted" energy, gave it to others. The old woman however, only had knowledge; this could be threatening. Increasingly she was defamed, and her knowledge truncated

---

[177] Ibid., 106.

[178] Dexter, *Whence the Goddesses,* 177.

by a discriminatory environment.[179]

Eve could be seen as a remnant of the Crone, since from the Judeo-Christian perspective, she is the cause of all death. In the fifth century C.E., a church council announced that it was heresy to say that death was natural rather than the result of Eve's disobedience.[180] As an Eve, every woman was "the devil's gateway" as announced by Tertullian. But Eve is really a very passive kind of Crone; she actually doesn't do the destroying. Christian theologians noted that the devil tempted Eve because she was weaker willed than Adam. Eve is a far cry from a Kali or a Lilith or a Medusa. Most of what she carries around is guilt, not wrath. And many women have taken on Eve's burden.

Medusa is exemplary of how Goddess' dark aspect became demonized in the patriarchal context. Marija Gimbutas points out that the earliest Greek gorgons were not terrifying symbols, but were portrayed with symbols of regeneration – bee wings and snakes as antennae.[181] Medusa with her serpent hair had been an ancient, recognized image of the divine female, a "ruling one."[182] In later mythology Perseus is celebrated as hero for defeating her by cutting off her head with its fearsome and so-called deadly gaze. There is no doubt that it is fearsome to look into the eye of the Divine; but patriarchal gods have carried the same characteristic, Yahweh for example, without threat of the same retribution. In the patriarchal context, is it really the gaze of the Female that is deadly? It is women who are the chronically gazed upon, whether as sex object or on a pedestal; She has been "kept an eye on." The beheading of Medusa, icon of Wisdom, may be understood as a story of dis-memberment of the Female Metaphor/Goddess.[183] The

---

179 Women were barred from education, yet at the same time denigrated as ignorant/foolish.

180 Walker, *The Woman's Encyclopedia of Myths and Secrets,* 290.

181 Gimbutas, *The Language of the Goddess,* xxiii.

182 Miriam Robbins Dexter, "The Ferocious and the Erotic: 'Beautiful' Medusa and the Neolithic Bird and Snake," *Journal of Feminist Studies in Religion*, Issue 26.1, 2010, 25-41: 25.

183 Just as the rape of Persephone – one who is Seed of Life/Redeemer/Eternal Thread – may be understood as a story of the dis-integration of the Female Metaphor/Goddess. These stories may be

hera's journey today is to go against the patriarchal injunction and look Medusa straight on, as philosopher Hélène Cixous suggests.[184] She is at first fearsome, but the Dark Goddess' fierceness may nurture life-giving strength in a woman, as she recognises the power within, and dares to take the journey into self-knowledge.

The Crone/Old One's realm is the waxing dark: She leads into the Void, the Space beyond and within all. Hers is the Underworld, the Place at the foundations of life, where form is broken-down, de-composed, dis-solved; where old ways are gone and the new can only be listened and felt for. It is necessary to re-value the Dark to understand Her. In earlier times the night was perceived as part of the day – the day was 'diurnal,' containing both light and dark aspects: and the night was alive. The day was reckoned from noon to noon, and midnight was centre position. What is called the "day," was understood to emerge out of the dark/night; as indeed all of manifestation does. For some religious traditions, the day still begins at dusk. The darkness of Goddess is a rich fertile Place, seething with potency. Her Darkness is where the new and undreamed of may be conceived, the "quantum vacuum" that physicists speak of: modern science enables a reapprehension of the "superessential darkness" of the Divine "out of which elementary particles emerge." Cosmologist Brian Swimme names this realm as "the all-nourishing abyss."[185]

The Crone's Dark Space is often symbolized in the Cauldron – a place of transformation, where the new is cooked up. She is the "Organizing Principle" that knows the recipe; She can be trusted to deliver from deep within. Her cauldron is not for mixing poison as has been told; it is a Cauldron of Creativity, frequently found at the bottom of deep fears, volcanic emotion, deep sadness. Within Her dark Space is found the essence for re-membering. As Patricia Reis reflects,

---

understood as records of the loss of an integrity that went before, just as Joseph Campbell notes was true of the story of the dismembering of Tiamat by Marduk, *The Power of Myth with Bill Moyers* (New York: Doubleday, 1988), 170.

[184] Hélène Cixous, "The Laugh of the Medusa" (Signs 1 no. 22): 885.

[185] Brian Swimme, in Dominic Flamiano, "A Conversation with Brian Swimme," *Original Blessing,* Nov/Dec 1997, 8-11: 10.

> Whenever I have felt the Dark Goddess' consciousness filling me there is always an accompanying dread. I know my life will never be the same. I know that I am being initiated into a new aspect of myself, a new part of my journey… And yet there is also a sureness, a firmness, a resoluteness, as in a re-solution.[186]

Often She is met through an accident, an illness, an emotional break – a tearing apart. Sometimes change may have been desired but the way unknown. Hers is an invitation to transformation. She can be *felt* in the need to exhale, to empty, in the release. She is felt in the ending of things, in the shed skins of old shapes and identities; as pain or joy or both. Her symbol is sometimes the sword; She cuts through illusion, and that vision is sometimes hard to bear. Hers is a fierce love, when there is love but something needs to change: She may be known in the "No more!"[187] and in the chaos of dismantling. Her gift is in the seed pod, the peeling bark, the pruned branches, the scissors cutting the thread – expressed in the image of the waning crescent moon, and felt, as that growing dark enters the eyes, comforting the sentience within, allowing new constellations to gestate.

I associate the Crone with the Dharma, the Truth-As-It-Is.[188] She sees all truth and can bear it, and Her compassion is without end. She allows the letting go of small self-limitations,[189] and is *She Who Returns Us to All.*

The Three Biological Shaping Powers and the Female Metaphor/Dea

Swimme and Berry note that biological life on planet Earth is

---

186 Reis, "The Dark Goddess," 82.
187 Quoting Bridget McKern, *Song of Hecate*; see Livingstone, *PaGaian Cosmology*, 267.
188 I acknowledge Joan Halifax, *Being With Dying,* for a broadened understanding of the Buddha, the Sangha and the Dharma – which I now associate with the three faces of the Female Metaphor.
189 Similar to the capacities of self-transcendence and self-dissolution of holons. See Willis Harman & Elisabet Sahtouris, *Biology Revisioned* (Berkeley: North Atlantic Books, 1998), 18.

shaped by "three fundamentally related, though distinct causes,"[190] and they reflect that these powers further illustrate the "root creativity" of the Universe that finds expression in the three qualities of Cosmogenesis.[191] These three shaping powers of the biosphere of life's journey here on Earth, are genetic mutation, natural selection and conscious choice/niche creation. I in turn find in their descriptions of these three biological shaping powers, a further articulation of the nature of the three faces of the Female Metaphor.

Swimme and Berry find in genetic mutation a biological illustration of differentiation – it is this power of mutation that gives rise to genetic variation. They describe it as a "pressure toward the future within each moment (that includes) a pressure for uniqueness,"[192] and I have come to identify Virgin energy – *the Urge To Be* – this way. They describe this dynamic with various words such as "*chance, random, stochastic* and *error*," finally summarizing the quality as *wild* – "a great beauty that seethes with intelligence, that is ever surprising and refreshing."[193] I associate such a description with the Virgin/Young One, particularly as She is celebrated at Beltaine/High Spring. I came to call this "the Poetry of genetic creation," which is an allusion to Thomas Berry's seventh principle of a functional cosmology, where he is stating the significance of the genetic coding process for life's expression and being.[194]

For Swimme and Berry, natural selection illustrates the dynamic of communion – it is this power that *sculpts* the diversity, *crafts* it. They describe natural selection as a

> dynamic of interrelatedness ... that presses, always and everywhere, for a deep intimacy of togetherness ... (deep into) the very structure of genes, body, mind.[195]

Swimme and Berry describe natural selection as a communal reality – a bonding process – wherein a species engages in finding its place in a

---

[190] Swimme and Berry, *The Universe Story*, 125.
[191] Ibid., 132.
[192] Ibid., 133.
[193] Ibid., 126-127.
[194] See Appendix A.
[195] Swimme and Berry, *The Universe Story*, 134.

biophysical community, and this seems similar to David Abram's understanding of it as a "reciprocal phenomenon:"[196] that is, a dialogue or conversation between the organism and the environment. The conventional and popular notion of the environment being "fixed," and to which the organism must conform was challenged by biologist Lynn Margulis in her groundbreaking research.[197] These descriptions of the flux between organism and Earth, as a co-creation of place, have deepened my understanding of the nature of the Mother face of the Female Metaphor – as *the Place to Be*; a Place that is a dynamic point of Interchange, a vibrant reciprocity, and that is celebrated particularly at the Winter and Summer Solstices. The Solstices are points of interchange between the light and the dark, the dark and the light, where one is seeded in the other. I came to call these Seasonal Moments, "Gateways" – places of Birth, either into form or into dissolution; they are points that celebrate life's transitions of birth and death, the holy Moments of the annual cycle that celebrate the interchange between the biological self (a singularity, be it species or individual) and existence (All-That-Is).

Swimme and Berry describe the third biological shaping power of niche creation or conscious choice as a biological illustration of the Cosmogenetic dynamic of autopoiesis.[198] Ordinarily, scientific accounts do not give niche creation/conscious choice as much importance as the other two biological shaping powers, but Swimme and Berry argue for its equal inclusion saying that at points of major evolutionary change, conscious choice becomes the primary explanation for the change. They call for more recognition of the self-organizing dynamics within all life

---

[196] Abram, *The Spell of the Sensuous*, 247.

[197] Margulis refers to the work of Russian scientist Vladimir Vernadsky and philosopher of science Karl Popper, saying that: "the activities of each organism lead to continuously changing environments. The oxygen we breathe, the humid atmosphere inside of which we live, and the mildly alkaline ocean waters in which the kelp and whales bathe are not determined by a physical universe run by mechanical laws; the surroundings are products of life interacting at the planet's surface." Cited in Connie Barlow, ed., *From Gaia to Selfish Genes: Selected Writings in the Life Sciences* (Massachusetts: MIT Press, 1994), 237.

[198] Swimme and Berry, *The Universe Story*, 132.

forms, of "behavior that can be interpreted as manifestations of memory, of discernment concerning questions of temperature and nutrient concentration, of a basic irreducible intelligence."[199] They express that even minimal powers of this kind have resulted in primal decisions on the part of organisms which have sent the biosphere into pathways forever characterized by those decisions. As a premise to their perception Swimme and Berry argue against the conventional notion of a "fixed environment" pointing out its limitations, stating rhetorically that a species always creates its own niche. They present the example of the horse and the bison who come from a common ancestor but are now very different forms of life – the different choices made by their primordial ancestors created two different worlds, with different selection pressures constellated for each, and these shaping the genetic diversity accordingly.[200] They describe this dynamic of niche creation as a felt "vision" or simple thrill wherein the creature responds to this inner attraction to pursue a particular path – much like the power of imagination draws the human. I associate this energy with the Crone/Old One, particularly as She is celebrated at Samhain/Deep Autumn, drawing forth the future, conceiving the new, from Her dark sentient depths. In the human this imaginative power is sometimes simply felt, sometimes "seen," always an act of will. I came to call this "the Poetry of trans-genetic creation" which is an allusion to Thomas Berry's ninth principle of a functional cosmology (See Appendix A), where he is stating the significance of human language – "cultural coding."

---

[199] Ibid.

[200] Ibid., 136-138.

CHAPTER 2
A PAGAIAN WHEEL OF THE YEAR

(Essentially a PaGaian Wheel of the Year celebrates Cosmogenesis – the unfolding of the Cosmos, none of which is separate from the unfolding of each unique place/region, and each unique being.) This creativity of Cosmogenesis is celebrated through Earth-Sun relationship as it may be expressed and experienced within any region of our Planet. PaGaian ceremony expresses this with *Triple Goddess* Poetry understood to be metaphor for the creative dynamics unfolding the Cosmos. At the heart of the Earth-Sun relationship is the dance of light and dark, the waxing and waning of both these qualities, as Earth orbits around our Mother Sun. This dance, which results in the manifestation of form and its dissolution (as expressed in the seasons), happens because of Earth's tilt in relationship with Sun: because this effects the intensity of regional receptivity to Sun's energy over the period of the yearly orbit. This tilt was something that happened in the evolution of our planet in its earliest of days – some four and a half billion years ago,[201] and then stabilised over time: and the climatic zones were further formed when Antarctica separated from Australia and South America, giving birth to the Antarctica Circumpolar Current, changing the circulation of water around all the continents ... just some thirty million years ago.[202]

Within the period since then, which also saw the advent of the earliest humans, Earth has gone through many climatic changes. It is likely that throughout those changes, the dance of light and dark in both hemispheres of the planet ... one always the opposite of the other – has been fairly stable and predictable. The resultant effect on flora and fauna regionally however has varied enormously depending on many other factors of Earth's ever-changing ecology: She is an alive Planet who continues to move and re-shape Herself. She is Herself subject to the cosmic dynamics of creativity – the forming and the dissolving and the re-emerging.

The earliest of humans must have received all this, 'observed' it in a very participatory way: that is, not as a Western industrialized or

---

[201] See Appendix C, *(6).
[202] See Appendix C, *(27).

dualistic mind would think of 'observation' today, but as kin with the events – identifying with their own experience of coming into being and passing away. There is evidence (as of this writing) to suggest that humans have expressed awareness of, and response to, the phenomenon of coming into being and passing away, as early as one hundred thousand years ago: ritual burial sites of that age have been found,[203] and more recently a site of *ongoing* ritual activity as old as seventy thousand years has been found.[204] The ceremonial celebration of the phenomenon of seasons probably came much later, particularly perhaps when humans began to settle down. These ceremonial celebrations of seasons apparently continued to reflect the awesomeness of existence as well as the marking of transitions of Sun back and forth across the horizon, which became an important method of telling the time for planting and harvesting and the movement of pastoral animals.

It seems that the resultant effect of the dance of light and dark on regional flora and fauna, has been fairly stable in recent millennia, the period during which many current Earth-based religious practices and expression arose. In our times, that is changing again. Humans have been, and are, a major part of bringing that change about. Ever since we migrated around the planet, humans have brought change, as any creature would: but humans have gained advantage and distinguished themselves by toolmaking, and increasingly domesticating/harnessing more of Earth's powers – fire being perhaps the first, and this also aided our migration. In recent times this harnessing/appropriating of Earth's powers became more intense and at the same time our numbers dramatically increased: and many of us filled with hubris, acting without consciousness or care of our relational context.

We are currently living in times when our planet is tangibly and visibly transforming: the seasons themselves as we have known them for millennia – as anyone's ancestors knew them – appear to be changing in most if not all regions of our Planet. Much predictable Poetry – sacred language – for expressing the quality of the Seasonal Moments will

---

[203] Swimme and Berry, *The Universe Story*, 152 and 274.
[204] Sheila Coulson, "World's Oldest Ritual Discovered – Worshipped the Python 70,000 years ago" (Science Daily, November 30, 2006). http://www.sciencedaily.com/releases/2006/11/061130081347.htm.

change, as regional flora changes, as the movement of animals and birds and sea creatures changes, as economies change.[205] In Earth's long story regional seasonal manifestation has changed before, but not so dramatically since the advent of much current Poetic expression for these transitions, as mixed as they are with layers of metaphor: that is, with layers of mythic eras, cultures and economies. We may learn and understand the traditional significance of much of the Poetry, the ceremony and symbol – the art – through which we could relate and converse with our place, as our ancestors may have done, but it will continue to evolve as all language must.

At the moment the *dance of dark and light* remains predictable, but much else is in a process of transformation. As we observe and sense our Place, our Habitat, as our ancestors also did, we can, and may yet still make Poetry of the dance of dark and light, of this quality of relationship with Sun, and how it may be manifesting in a particular region and its significance for the inhabitants: we may still find Poetic expression with which to celebrate the sacred journey that we make every day around Mother Sun, our Source of life and energy. It has been characteristic of humans for at least several tens of thousands of years, to create ceremony and symbol by which we could relate with the creative dynamics of our place, and perhaps it was initially a method of *coming to terms* with these dynamics – with the apparently uniquely human *awareness* of coming into being and passing away.[206] Our need for sacred ceremony of relationship with our place, can only be more dire in these times, as we are witness to, and aware of, ever larger dimensions to the beauty and terror of being.

With all this in mind, I will proceed with sharing the PaGaian ceremonial celebration of Earth's annual journey around Sun, as it eventuated at my place in the Blue Mountains of Australia in the decades before and after the turning into the twenty-first century of this Common Era.

---

[205] I do not wish to make light of the pain of these changes, for humans and all creatures, but change is happening and all will adapt as before – in action and story.

[206] As Swimme and Berry suggest in *The Universe Story*, 152-153.

Dark and Light in the Wheel of the Year

An open community of participants celebrated eight Seasonal Moments throughout the year at my sacred place, according to what I consider to be the pre-Celtic Wheel of the Year as I learned it originally from Starhawk's tradition.[207] These eight Seasonal Moments consist of: two Solstices – Winter and Summer, two Equinoxes – Autumn and Spring, and four cross-quarter days, that is, the meridian points of each quarter of the year. There is a simple division into a light half of the year during which the light is waxing – Winter through Spring to Summer, and a dark half of the year during which the dark is waxing – Summer through Autumn to Winter. There is another division of the year into one half when the hours of light in a day are longer than the hours of dark – Spring Equinox through the Summer to Autumn Equinox, and the other half being when the hours of dark in a day are longer than the hours of light – Autumn Equinox through Winter to Spring Equinox.
On its simplest level, the light that is 'born' at the Winter Solstice waxes through the Spring, at first being young and tender, before growing strong and coming into balance with the dark at Spring Equinox, then waxing into the strength and passion of High Spring (Beltaine) and Summer. At the Summer Solstice, the dark that is 'born' waxes through the Late Summer/Autumn, gradually spreading its cloak, before growing strong and coming into balance with the light at Autumn Equinox, then waxing on into the dark transformation of Deep Autumn (Samhain) and Winter. The diagrams on the pages that follow are a map of the Wheel of the Year – one for the Southern Hemisphere and one for the Northern Hemisphere, with the PaGaian perspective and story included. The dates given are based on the actual astronomical date, which varies.[208]

---

[207] See Starhawk, *The Spiral Dance.*

[208] The actual astronomical date of the Seasonal Moment varies from year to year because it is the meridian point of the quarter (the "cross-quarter day"). All global times may be found at pagaian.org. In terms of representing the dance of dark and light on the diagram, I find it best to express it with the yin-yang arrangement, which is apparently based on Earth-based Chinese tradition. Annabelle Solomon notes that the South should really be at the top of the page for the Southern Hemisphere perspective. She does this in her diagram of the Wheel of the Year, *The*

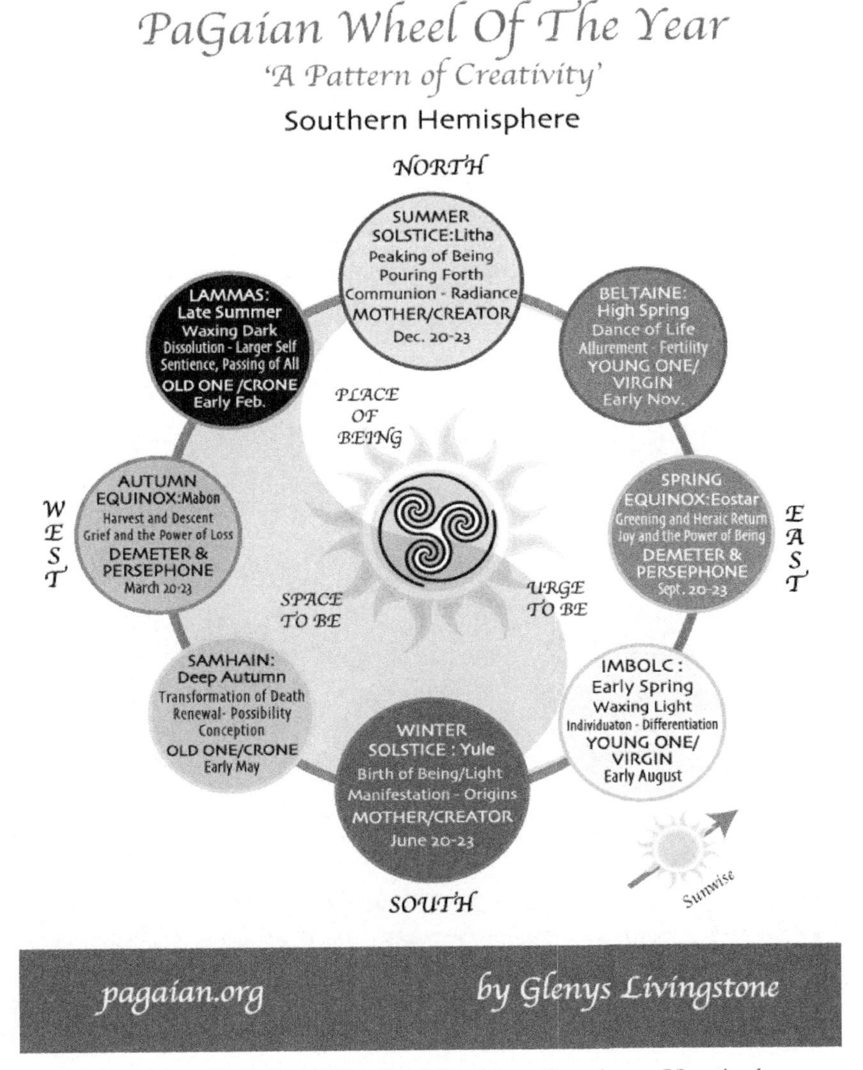

[Figure 6] PaGaian Wheel of the Year Southern Hemisphere

Wheel of the Year: Seasons of the Soul in Quilts (Winmalee NSW: Pentacle Books, 1997), 105, and see Livingstone, *PaGaian Cosmology*, Appendix E: https://pagaian.org/book/appendix-e/.

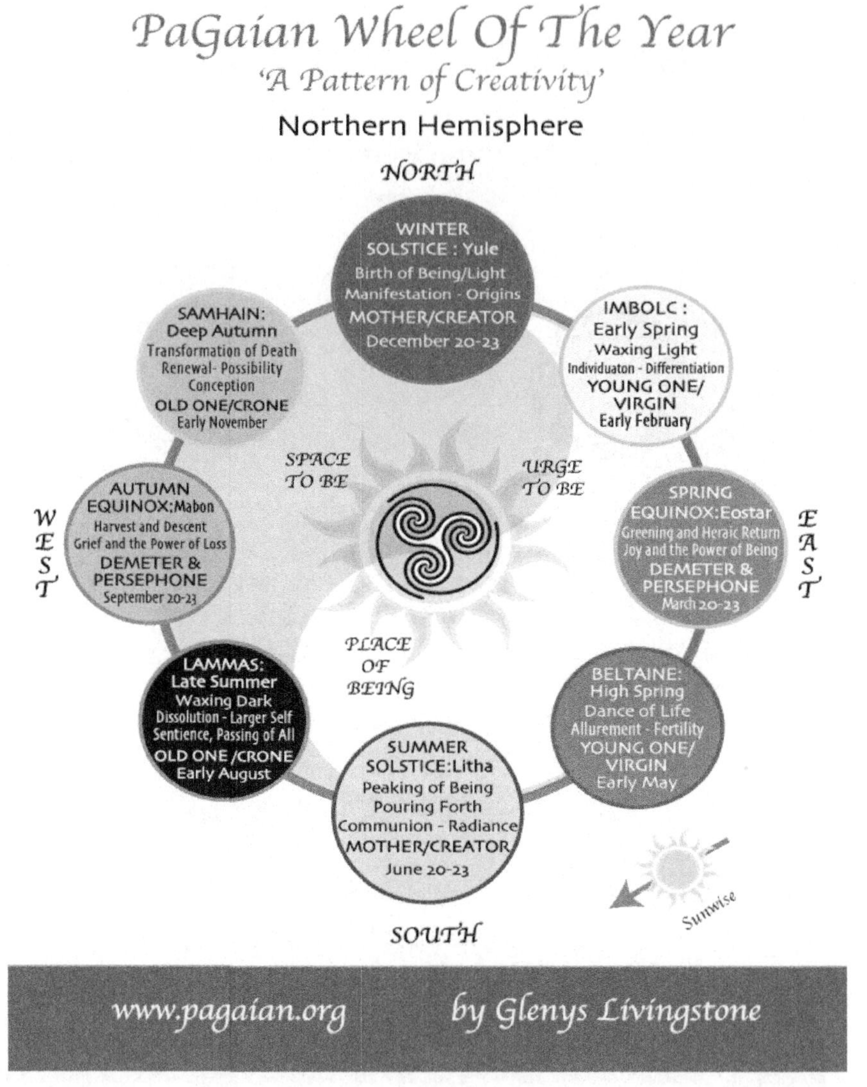

[Figure 7] PaGaian Wheel of the Year Northern Hemisphere

I commonly think of the light part of the cycle as acknowledging and celebrating the *manifest* reality, and the dark part of the cycle as acknowledging and celebrating the *manifesting* reality. I take these terms from David Abram's explanation of Benjamin Lee Whorf's work in analyzing the Hopi language, because it seems from Abram's interpretation of Whorf's work that the Hopi had a sense of space and time that was similar to that invoked by practice of the Seasonal Wheel of the Year. Abram says:

> While Whorf did not find separable notions of space and time among the Hopi, he did discern, in the Hopi language, a distinction between two basic modalities of existence, which he terms the 'manifested' and the 'manifesting.'[209]

Abram summarizes the meaning of these terms with:

> The 'manifested,' … is that aspect of phenomena already evident to our senses, while the 'manifesting' is that which is not yet explicit, not yet present to the senses, but which is assumed to be psychologically gathering itself toward manifestation within the depths of all sensible phenomena.[210]

It is easy enough for the average modern Westernised and Christianised mind to associate light with manifestation; the birth of light at the point of Winter Solstice and its waxing through to the fullness of Summer Solstice is fairly easily taken on as cause for celebration. However, the celebration of the dark is quite another thing; the average modern Westernised and Christianised mind[211] finds it harder to

---

[209] Abram, *The Spell of the Sensuous*, 191. At first, I used the terms "manifest" and "unmanifest," which altered the sense slightly, but it seemed clearer to me given my cultural context of lack of familiarity with the unseen. I now use the terms "manifested" and "manifesting" if the meaning to others is clear, but it often may require qualifying.

[210] Ibid., 192.

[211] One might add 'patriarchal' mind to this alienated, dualistic frame of reference; and this state of mind is clearly cross-cultural.

comprehend, having metaphorized the dark as dead-end, bad, even sordid; and thus largely in a state of denial that it is part of life. Within a linear time frame, where the dark is disconnected from the cycle, the dark is no longer a space for transformation, for 'manifesting'. Participation in the Wheel of the Year process re-enables the positive and creative **sense** of the dark. With experience of the Wheel's cycle (especially with a Southern Hemispheric mind), one may come to always be aware of the polar opposite Seasonal Moment, in the midst of each particular seasonal celebration: that is, to be aware of the presence of the 'manifested' in the 'manifesting,' and vice versa.[212] For example, at Lammas (Late Summer/Early Autumn) where the individual self ceremoniously assents to giving over to the dark, there is a memory of Imbolc (Early Spring), when differentiated unique self was celebrated and nurtured; at Lammas, one may become aware that 'I' am simply *returning* the manifested to the Source or the "*heart* ... behind and within all the forms and appearances of nature."[213].

The practice of ceremonial celebration of the Seasonal Wheel of the Year invokes and nurtures a sense of space and time wherein

> ... one's own feeling, thinking, desiring are a part of, and hence participant with, this collective desiring and preparing implicit in all things – from the emergence and fruition of corn, to the formation of clouds and the bestowal of rain. Indeed, human intention, especially when concentrated by communal ceremony and prayer, contributes directly to the becoming manifested of such phenomena.[214]

---

[212] This awareness/memory of the polar opposite Seasonal Moment within any current one is enhanced by awareness of the other Hemisphere's polar opposite inclination/Moment; that is, by a global/PaGaian perspective. It has been easier for inhabitants of the Southern Hemisphere to appreciate this.

[213] Abram, *The Spell of the Sensuous*, 192, quoting Benjamin Lee Whorf in Dennis Tedlock and Barbara Tedlock, eds., *Teachings from the American Earth* (NY: Liveright, 1975), 122.

[214] Abram, *The Spell of the Sensuous*, 192.

Local-particular and Cosmic at the same time

Participation in the Wheel process, particularly when practised as a whole year-long experience and over the period of years, re-identifies one's small self with the Larger Gaia-Self. It is the experience of many indigenous cultures that their communal ceremony and prayer, along with their daily activities, participate "in acts that evoke the ongoing creation of the cosmos."[215] Increasingly, as I practiced this Wheel of the Year process, I came to understand how we create the Cosmos, whether conscious or not. I did not know this when I began; this awareness grew in the practice of identifying with the Creativity of Gaia Herself, through the cycle of the 'manifest' and the 'manifesting,' the light and the dark, the differentiation and the transformation. Gradually I came to understand (and continue to understand more deeply) how these seasonal ceremonies are a response to awakened relationship with Cosmos – thus in some sense, the celebrations become a

> *responsibility* to the cosmos ... to know grace, to know as intimately as possible the mysterious interrelatedness and spiritual powers that infuse being and to live our lives accordingly.[216]

Perhaps the central/essential significance of the Seasonal Moments (Sabbats) is that they are points of expression of relationship with Gaia, who is a Phenomena of storied events. These "events" are not accomplished and "located in some finished past" but are "the very depth of the experiential present," as Abram describes this sense when describing his understanding of the Aboriginal Australian notion of Dreamtime (or "Alcheringa").[217] Abram understands the "Dreamtime" /Alcheringa to refer to the "implicit life" of a place, the storied events that "crouch within" a place, that the human may rejuvenate with

---

215 Spretnak, *States of Grace,* 95.
216 Ibid., 100.
217 *The Spell of the Sensuous*, 193. "Alcheringa" is the Indigenous term, and there is some dissatisfaction with the translation of the term as "Dreamtime" insofar as it may imply something "not real." For more: https://en.wikipedia.org/wiki/Dreamtime.

"en-*chant*-ment" and action.[218] The human is thus enchanted and rejuvenated simultaneously – human and Habitat know relationship, intimacy. This seems resonant with what I have called "sacred awareness" of the Universe we live in, that may be conjured by the ceremonies of the Wheel of the Year. After celebrating the seasonal ceremonies for a period of time I came to have a clear sense, at each Seasonal Moment, as I/we prepared to celebrate the ceremony, of the uniqueness and depth of this space-time moment in the history of the Universe; I understand this as "sacred awareness." It is a sense of the deep time and space of the moment, and that it is significant Cosmically.[219] The moment becomes a Moment, as actually all moments are; this sense then has often noticeably carried over into my life at other moments.

The Poetry of each Seasonal Moment, as humans have scribed it over eons, has its roots in relationship of Earth with Sun primarily; then in how this has affected Earth's response, and our response in the growing of our food and in the tending of animals (both pre and post-domestication); then in how it has affected human understanding of the Mystery at the Heart of existence – how we have storied it, metaphorized it, how we have come to understand life and death through the participatory observing of this creative process. The Poetry then expresses how humans have come to understand personal and collective stories – the search for the harmony within our stories' pain and ecstasy. All these layers of story over time, and even varying within a place, and a culture, are intermingled, making a web that is complex, and ever more so; but each seasonal point does retain a particular moment of relationship, a particular 'slice' of the *Whole Creative Cycle* of light and dark essentially – of 'manifest' and 'manifesting' – that invites celebration. And gradually, the place in which one celebrates this Great Story … the place which is local-particular and cosmic at the same time, becomes **the** sacred site, as author Caitlin Matthews, speaks of it when she describes the experience of consciously joining Earth in Her circumambulation of

---

[218] Ibid.

[219] Even though the actual Cosmic significance does lie well beyond anyone's comprehension.

Sun.[220] The ceremonial celebration of each Seasonal Moment/Sabbat – the moment in time, the particular 'slice' of the Whole Creative Cycle of light and dark, may embed one in a growing *sense* – felt knowledge – of being at the Centre and Origins of All.

Some of the words/poetry in the ceremonies as I have celebrated them are taken directly from the traditional expressions/stories that Starhawk articulates in her classic book *The Spiral Dance.* I have also added much of my own understanding and storied in the particular emphasis of the Triple Goddess Triple Spiral/Cosmogenesis. Starhawk has always invited adaptation of the ceremonies as need, place and circumstance predicated. Her version includes a much stronger presence of the God, which is a later (Celtic) adaptation of a story originally based in the phases of the Triple Goddess.[221] There is not any right and wrong way to tell the story of the Wheel of the Year, to speak the great creative process; there seems to be as many interpretations as there are groups of people and places. These Sabbats/festivals have taken different names in different ethnic groups over the centuries, and their significance altered slightly according to the culture. As one example, Emma Restall Orr gives various names of the festivals and some of the different emphases,[222] and from a reading of Shirley Toulson's explanation of the Celtic Year[223] it seems that the ancestors moved dates around to suit their perceptions of the light and dark, and the stories they wanted to tell. Starhawk points out that the story will be different for specific places and climates, and that "the way we celebrate the seasonal festivals also changes over time."[224] She explains the celebrations as she has described them in *The*

---

220 Matthews, *The Celtic Spirit*, 339.

221 Susan Gray, *The Woman's Book of Runes* (New York: Barnes and Noble, 1999), 17-18. Adam McLean, *The Four Fire Festivals* (Edinburgh: Megalithic Research Publications, 1979), and Durdin-Robertson, *The Year of the Goddess*, both also note this earlier (pre-Celtic) Goddess version of the Wheel of the Year and detail its celebration to some extent.

222 Emma Restall Orr, *Spirits of the Sacred Grove* (London: Thorsons, 1998), 234-235.

223 Shirley Toulson, *The Celtic Year*. Dorset: Element Books Ltd., 1993.

224 Starhawk, *The Spiral Dance*, 257.

Spiral Dance:

> Some of these rituals – or some aspects of them – we still do very much as is described here. For some festivals, new traditions have evolved that are repeated year after year. For others, each year's ritual is different. Some festivals have specific children's rituals involved; others don't as yet. The opening invocations reflect the old imagery based on heterosexual polarity. Please feel free to change or adapt them. I may rewrite them someday, after a few more turns around the wheel.[225]

Caitlin and John Matthews, scholars and practitioners of the "Western Way," point out that the seasonal festivals have much deeper significance than the exoteric celebration of them makes apparent.[226] They note how the forms and names of the festivals have changed but declare that "the inner protective energy remains the same," and that "they are the times when the way to the Otherworld stands open."[227] The festivals, they say, can become "hidden doorways in your life" for vital inner energy and "fostering sensitivity to the important psychic tides which energize the world."[228] It has been so, in my experience, and also in the experience of the participants who spoke of it.

In PaGaian Cosmology, I have adapted the Wheel as a way of celebrating the Female Metaphor and also as a way of celebrating Cosmogenesis, the Creativity that is present really/actually in every moment, but for which the Seasonal Moments provide a pattern/Poetry over the period of a year – in time and place. The pattern that I unfold is a way in which the three different phases/characteristics interplay. In fact, the way in which they interplay seems infinite, the way they interrelate is deeply complex. I think it is possible to find many ways to celebrate them. There is nothing concrete about the chosen story/ Poetry, nor about each of the scripts presented here, just as there is

---

[225] Ibid.

[226] Caitlin and John Matthews, *The Western Way* (London: Penguin, 1994), 47.

[227] Ibid., 47-48.

[228] Ibid., 48.

nothing concrete about the Place of Being – it (She) is always relational, a *Dynamic Interchange.* Whilst being grounded in the "Real," the Poetry chosen for expression is therefore at the same time, a potentially infinite expression, according to the heart and mind of the storyteller.

Swimme and Berry call for "a more symbolic language ... to enter into the subjective depth of things, to understand both the qualitative differences and the multivalent aspects of every reality."[229] The Wheel of the Year may be such a multivalent language, I believe. It can speak any number and layers of experiences in the dynamic of the waxing into fullness and the waning into emptiness – in our individual lives, in the life of a group, in the life of Earth, in the evolution of the Universe. It happens in time, it is an objective event, but its interpretation is multivalent. I use the Wheel of the Year, and its traditional base, to spell out a cosmology, a Poiesis, as I perceive it.

Following is a summary description of the Wheel, as one Seasonal Moment kaleidoscopes into the next. It poetically aligns Cosmogenesis with the Wheel of the Year, as will be described in more detail in ensuing chapters. This alignment is drawn into the Poetry of the ceremonies, as is further described in chapter 3. The traditional dates for both hemispheres are given below, with the Southern Hemisphere designated "S.H." and the Northern Hemisphere designated as "N.H.," and as noted earlier the actual astronomical dates anywhere on the globe do vary from year to year.[230]

Samhain/Deep Autumn – April 30th S.H., October 31st N.H.

Samhain traditionally marks the New Year in the practice of this cosmology because it is the meridian point of the darkest phase of the

---

229 Swimme and Berry, *The Universe Story,* 258.

230 See endnote 8 of this chapter. Also of note, Vivianne Crowley, *Celtic Wisdom: Seasonal Rituals and Festivals* (New York: Sterling, 1998), 104, points out that prior to reliance on calendar and clock time, some seasonal festival dates were not fixed, but were timed according to harvest and weather and locality. Caitlin and John Matthews, *The Western Way*, 47, say that: "In inner terms, the right time is more important than the right date. Those living in the Southern Hemisphere are already well aware of this."

year, wherein the new is understood to be conceived. There is argument that Winter Solstice marks the New Year, and it may be felt to be so. Starhawk places Winter Solstice first in her chapter on the Wheel of the Year in *The Spiral Dance*, and Samhain at the end of that chapter, although she notes this ending of the year as the New Year and adds: "… so we end in the beginning, as we should, and the Wheel turns on."[231] Wherever the New Year is marked exactly, it is understood that the New is conceived in the darkest part of the annual cycle, of which Samhain Moment is meridian: this is the timeless and magic moment between the old and the new. 'Samhain' means literally "Summer's end," when last Summer's growth falls away, and Winter darkness embraces the land. (I like to begin here, because I like so much to celebrate conception, the dreaming, the imagining, as the beginning of form, some time before birth when the form becomes visible. Much like any New Year, when "re-solutions" are made, all is possible; one could decide anything and will it to be so. This is of course true of every moment, and this celebration can remind one of that.)

I have adapted Starhawk's traditional telling of the Samhain ceremonial moment[232] for PaGaian Cosmology, and tell it in the following way:

> This is the time when we recognize that the veil is thin that divides the worlds. It is the New Year in the time of the year's death – the passing of old growth. The leaves are turning and falling, the dark continues to grow, the days are getting shorter and colder. Earth's tilt continues to move us away from the Sun.
>
> The story of old tells us that on this night, between the dead and the born, between the old and the new, all is possible; that we travel in the Womb of the Mother, the Dark Shining One within, from which all pours forth, and that we are the seeds of our own rebirth. The gates of life and death are opened: the dead are remembered, the Not-Yet[233] is conceived. We meet in time out of

231 Starhawk, *The Spiral Dance*, 209.

232 Ibid., 209-210.

233 A term used by Brian Swimme, *The Earth's Imagination* (DVD series 1998), Program 8, "The Surprise of Cosmogenesis."

time, everywhere and nowhere, here and there ... to transform the old into the new in our own bodyminds.

Samhain is a profound celebration of the Void – the Void before time, the Space between one exhalation and the next inhalation, the All-Nourishing Abyss, the Sea of Generosity[234] from which all pours forth, the quantum vacuum. It is regarded as the time for remembering the ancestors – those who have gone before us. From the point of view of Gaia/All-That-Is, death is a transformation; Samhain is a time for remembering this, and for being done with the old and conceiving the new. It is a time for recalling the many changes each participant has come through in their particular lives. It can recall the many cultural changes of our human story, the many evolutionary changes – Gaia's transformations, which are also ours – anywhere in the spectrum of thirteen point seven billion years. We can remember how old we really are, and we can remember that we are yet "much More,"[235] as is stated in the PaGaian ceremony – personally and collectively. We may articulate some of these conceptions. This is the autopoietic quality of Cosmogenesis, of the unfolding of the Cosmos.

Samhain is a celebration of the Old One/Crone's process of the transformation of Death. She is the Old One who remembers, and from whose sentient depths the new is drawn forth. The imagined conceptions will continue to gestate in the fertility of the Old One's growing dark. At Samhain, Her face has begun to move into Mother – the Womb of Winter Solstice.

Winter Solstice/Yule – June 20–23 S.H., December 20-23 N.H.

This is one of the easiest of Earth's holy days for people of recent centuries in general to relate to, particularly in Christianized cultures

---

[234] The Sufi poet Iraqi uses this term for Origins. See William C. Chittick and Peter Lamborn Wilson, trans., *Fakhruddin 'Iraqi: Divine Flashes* (London: SPCK, 1982), 76. Swimme also speaks of how "all being has gushed forth because Ultimate Generosity retains no thing," *The Universe is a Green Dragon*, 146.

[235] This is a term Jean Houston has used to describe human and evolutionary potential.

where it has been celebrated as "Christmas," and in European cultures since the Middle Ages. The Winter Solstice marks the stillpoint in the depths of Winter, when Earth's tilt and orbit cause the Sun to begin its return. It may have been among the first Earth-Sun events that the ancients noticed: it is the most obvious and dramatic (and hoped for) for some regions of Earth particularly, yet it has been especially marked cross-culturally. It is this Seasonal Moment for which the ancients in Ireland apparently built "Newgrange" (as it has been named, though its Indigenous name is Brú-na-Bóinne): this monument was thought to be a burial mound or temple-tomb until recently, but evidence now makes clear that it in fact celebrates ongoing Earth-Sun creativity.[236] The inner chamber wall is carved with the Triple Spiral, which at this Seasonal Moment is illuminated for some seventeen minutes by the rising sun; thus expressing the significance of the Winter Solstice.

In this tradition since Celtic times, and in many other cultural traditions, this Moment has been celebrated as the birth of the God (often identified as the Sun since the rising of the gods): yet for most of humanity's history, the Sun was understood as Mother, not as a male principle,[237] so the story may vary accordingly. The PaGaian ceremonial statement of purpose is a variation of Starhawk's,[238] emphasizing that what is born, is within each one – the 'Divine' is not 'out there,' we are each Created and Creator:

> This is the time of Winter Solstice in our Hemisphere. Earth's tilt leans us away from the Sun to the furthest point at this time in our annual orbit. This is for us, the time when the dark part of the day is longest – darkness reaches Her fullness, She spreads her cloak, and yet gives way, and moves back into light. The breath of nature in our part of the world is suspended. She rests. We wait ... within the Cauldron, the Dark Space, for the transformation.
>
> The stories of Old tell of the Great Mother giving birth to the Divine Child on this night. This Divine Child is the new being in

---

[236] See Brennan, *The Stones of Time.*

[237] See Patricia Monaghan, *O Mother Sun! A New View of the Cosmic Feminine* (Freedom CA: Crossing Press, 1994).

[238] Starhawk, *The Spiral Dance*, 198.

you, in me ... is the bringer of hope, the light in the darkness, the evergreen tree, the centre which is also the circumference – All of Manifestation. The Divine Child being born is the Miracle of Being, and the Unimaginable More that we are becoming.

Winter Solstice is the time for rejoicing in the awesome miracle of manifestation – at the beginning of time, and in every moment. It is a celebration of the Primeval Fireball – the Original Big Birth, as well as the actual birth of our Sun from the Grandmother supernova, and the birth of the first cell, and our own personal manifestation; and it is the time for the lighting of candles, and expressing what we will birth in ourselves in the coming year. It may be a moment for recalling the Great Turning of these times, as Joanna Macy has named the great transition being made[239] – the hope we might hold for the future.

Winter Solstice is a celebration of the Mother aspect of Creativity, the ripening of Her Darkness into the awesome act of creation of form, the Web of Life, the Field of Being. It is a celebration of Communion, a point of interchange from the *manifesting* into the *manifest*; it is a time for feasting and community and experiencing this joyful essence of existence. At this point in the Wheel She is the Alpha, and at the Summer Solstice She will be the Omega – both Gateways, points of interchange, when dark and light turn. At this Winter Gateway, the Crone's face passes through the Mother to the Virgin/Young One. The process of the three Seasonal Moments of Samhain, Winter Solstice and Imbolc, as a triad, may be felt as the three faces of Cosmogenesis in the movement towards form.

Imbolc/Early Spring – August 1st S.H., February 1st N.H.

"Imbolc," as it may be named, is the meridian point of the new quarter, and is quintessentially the celebration of the new, which is understood to be continually emerging. It is the first celebration proper, of the light part of the cycle, and as such, it recognizes the vulnerability, the fragility of that new light or being, of those first tendrils of green, that

---

239 Joanna Macy and Molly Young Brown, *Coming Back to Life* (Gabriola Island, Canada: New Society Publishers, 1998), 17.

new self.[240] It is especially dedicated to the Virgin quality, inviolable in Her commitment to Being, Her 'yes' to life; She is traditionally invoked as Brigid in this cosmology, who is the tender of the sacred Flame of Life.

Using Starhawk's words,[241] in combination with my own emphasis, I state the seasonal purpose thus:

> This is the season of the waxing light. Earth's tilt is taking us back towards the Sun. The seed of light born at the Winter Solstice begins to manifest, and we who were midwives to this infant Flame now see it grow strong as the light part of the day grows visibly longer. This is the time when we celebrate individuation: how we each become uniquely ourselves. It is the time when we celebrate beginnings … the first tendrils of green emerging tentatively from the seed. We meet to share the light of inspiration and creative intentions, which will grow with the growing year.
>
> This is the Feast of the Virgin – Brigid, She who tends the Flame of Being; Artemis, She who midwifes body and soul. She is deeply committed to the Creative Urge, to manifestation, deeply committed to Self, and bringing forth the New, the Promised One. She is uncompromised, unswerving, noble, true, a warrior of spirit. She will protect the stirrings of Life.

For women particularly, the Imbolc process/ceremony can be an important integrating expression and movement, accustomed as many women are, to fragmentation in relationship – giving themselves away too easily. This seasonal celebration of movement into form, individuation/differentiation, yet with integrity/wholeness, especially invoking She-who-is-unto-Herself, can be a significant dedication. It is a 'Bridal' commitment to Being, in the original Brigid-ine sense.[242] All

---

[240] The cold has usually been increasing since Winter Solstice, so the inhabitants are often needing to "rug up" and seek the warmth and nourishment of the light.

[241] Starhawk, *The Spiral Dance*, 202.

[242] "Bride" is another name for Brigid, Great Celtic Goddess from pre-Christian times. See Crowley, *Celtic Wisdom,* 57.

including men may find this celebration of Brigid to be an integrating invocation within themselves – at last an opportunity to identify with "Her," to Whom we/they belong. For any being – identified as female or male or both/neither – it may be a statement of taking up courage to be, and a celebration of the individual quest.

The lighting of candles and a central flame is again a big part of this Seasonal Moment as it was at Winter Solstice, this time recognizing that each self is a 'Promise of Life.'[243] Each individual Promise is identified with Gaia Herself, with "the beauty of the green earth and the white moon among the stars and the mystery of the waters."[244] It is a time for purification; that is, for recognizing what it is in you that inhibits the spark, the growth, the power to be, and what enhances it; then for making a commitment to the tending of this Self. This Earth holy day particularly celebrates differentiation, diversity, the multiform beauty of Gaia, all of which is indeed brought to us through the many challenges that Gaia Herself has encountered as She has developed; and our individual lives are no different. The challenges we have encountered and midwifed ourselves through, may add to our complexity, strength and beauty. The Virgin is that aspect that finds the "yes" to being – the capacity to reach beyond the complete awesomeness of being, both personally and collectively. This aspect finds the "yes" to loving the self/Self beyond all failings and is able to step into the 'Power of Life' – so She moves into the balance of Eostar/Spring Equinox.

Spring Equinox/Eostar – September 20–23 (S.H.), March 20-23 (N.H.)

As the light continues to grow, it comes into balance with the dark. Spring Equinox, or "Eostar" as it may be named, is one of two points in the year when the Sun is equidistant between North and South, creating this light and dark balance. Yet the trend at this Equinox is toward increasing light – longer hours of light in the day; Earth in this region is still tilting further toward the Sun. Traditionally it is the joyful celebration

243 The term is derived from Starhawk's 'Child of Promise,' *The Spiral Dance,* 236.

244 Ibid., 102-103, quoting Doreen Valiente's "The Charge of the Goddess."

of Persephone's return from the underworld; this is when the balance tips – the certainty of the return of light is assured, the darkness has been navigated successfully. As is said in the PaGaian seasonal ceremonial statement, rewording Starhawk slightly:

> This is the time of Spring's return. Warmth and growth may be sensed in the land, the flowers can be smelt in the air. The young light that we celebrated at Imbolc, has grown strong and come to balance with the dark. Life bursts forth with new strength. The story of Old tells us that Persephone, beloved Daughter, returns from Her journey to the Underworld – Demeter stretches out Her arms – to receive and rejoice. The Beloved One, the Lost One, returns with new Wisdom from the depths.
>
> We may step into a new harmony. Where we step, wild flowers may appear; where we dance, despair may turn to hope, sorrow to joy, want to abundance. May our hearts open with the Spring.[245]

The patriarchal version of Persephone's story is that She is abducted and raped by Hades.[246] I think it is particularly important that this myth be re-storied. As it has been known in patriarchal times, it is an account of what did happen historically in the human story: that is, the hera was taken and raped, we were diminished. However, in the oldest tale, Persephone *voluntarily* descends to the underworld – she is not forced.[247] She has the wisdom of Goddess, who understands the fertility of the Dark terrain, who holds within Herself the Mystery of life and death. In this old account, Persephone journeys seeking self-knowledge and compassion, retaining Her integrity and sovereignty.[248] She is thus

---

245 Ibid., 204.

246 Though other variations are also possible, and sometimes developed by some Pagans: for example, that Persephone falls in love with Hades and runs off with him.

247 See Spretnak, *Lost Goddesses of Early Greece*, 105 -118.

248 This retention of Her integrity does not mean that She does not really lose Herself as the Hera/Hero must in the Descent to the Underworld. It simply means that She remains a Wisdom figure – the Hera, just as for instance, Inanna of Sumer retains Her identity as

restored to Her former grace, and to the gaining of a sense of Her full participation in the mystery and adventure of Life.

Persephone's return is the certain return of manifest Creativity. She brings with Her, knowledge of the Depths (autopoiesis), from whence springs all Creativity. Persephone's journey is about becoming familiar with the inner realms in herself, and falling in love with these depths: as well as Her role as Restorer of Beauty/Redeemer. In the Creation story of the Faery tradition, all manifestation springs forth from Goddess falling in love with Her reflection in the curved mirror of black space.[249] The ancients understood that the essence of Creative Power springs from self-love, known and seen only completely in the Dark.

I story this celebration as a "Stepping into Power" (the Joy and Power of Being), identifying ourselves as Heras, rejoicing in how we have made it through so much, having faced our fears, the chthonic, and our demise (in its various forms) – personally and collectively. It is a time to welcome back that which was lost, and step forward into the growing light, to fly. Eostar is the time for enjoying the fruits of the descent, of the journey taken into the darkness. It is a point of balance of the three faces of Goddess – Persephone representing both the Wise One from the depths and the newly Emerged, being embraced by the Mother, rejoicing and affirming the harmony of All. It is the three aspects of Cosmogenesis in "a fecund balance of tensions."[250]

Earth is perfectly poised in this balance for a moment, before She tips into the increasing fertility of Spring. The freedom of empowerment, the exhilaration of the full flight of Being, brings with it increasing passion for Life. Allurement awakens, desire reaches for "More" (promised at Samhain), for fullness; it is the wild, untamed nature of the Virgin who would give Herself to the ecstatic Dance of Life. This is the energy of Beltaine.

Queen/Sovereign in Her Journey to the Great Below where She becomes a rotten piece of meat.

249 Starhawk, *The Spiral Dance*, 41.

250 Swimme and Berry, *The Universe Story,* 54.

Beltaine/High Spring – November 1st (S.H.), May 1st (N.H.)

Earth's holy day of Beltaine marks the meridian point of the lightest phase in the cycle – it may be named High Spring; the time when the light part of the day is longer and continuing to grow longer than the dark part of the day. Beltaine is polar opposite Samhain on the Wheel of the Year, when the dark was still climaxing.

Based on Starhawk's telling of it,[251] but mostly in my own composition, I express the seasonal ceremonial purpose thus:

> This is the time of Beltaine, when the light part of the day is longer and continues to grow longer than the dark part of the day. Light is waxing towards fullness. In our region of the world, Earth continues to tilt us further toward the Sun – the Source of Her pleasure, life and ecstasy. This is the time when sweet desire for life is met. The fruiting begins. It is the celebration of allurement … Holy Lust ... that which holds all things in form and allows the dance of life.
>
> The ancients called this Holy Lust, this primordial Essence "Ishtar," "Venus," "Aphrodite" … they sang of Her:
>
> 'For all things are from you.
> Who unites the cosmos.
> You will the three-fold fates.
> You bring forth all things.
> Whatever is in the Heavens.
> And in the much fruitful earth
> And in the deep sea.'
>
> (Let us celebrate our erotic nature, which brings forth all things.)

(Beltaine is an opportunity to recognize and express our desire for life, which we feel in so many ways; and to recognize that it is a holy Desire. On an elemental level, there is our desire for Air, Water, the warmth of Fire, and to be of use to Earth. There is an essential longing, sometimes nameless, sometimes constellated, experienced physically, that may be

---

[251] Starhawk, *The Spiral Dance*, 205.

recognized as the Desire of the Universe Herself, desiring in us.[252] We may remember that we are united in this desire with each other, with all who have gone before us, and with all who come after us – all who dance the Dance of Life. Beltaine is a time for dancing and weaving into our lives, our heart's desires; traditionally the dance is done with participants holding ribbons attached to a pole/tree.[253] There is also the exhilarating tradition of leaping the flames, exclaiming what one wills to leave behind – it may be understood as letting the Flame of Love burn away the perceived blocks to one's desires. Beltaine is a time for assenting to the pace of the Dance of Life, with only Passion as the guide for where to place one's feet; much like the dancing Goddesses and Gods of many spiritual traditions. It is being with Life and its intense fertility, in the moment.

---

[252] Swimme, *Canticle to the Cosmos,* video 2, "The Primeval Fireball," and video 10, "The Timing of Creativity."
[253] I have named this a "Novapole," for our Southern Hemisphere, whereas in the Northern Hemisphere it is known as a "Maypole."

[Figure 8] Beltaine Tree, MoonCourt

One of the shaping powers of life is a wild energy, that Swimme and Berry associate with the causal factor of genetic mutation, describing that "Genetic mutation refers to spontaneous differentiations taking place at life's root."[254] Wild energy is also associated with Artemis in Her Virgin aspect;[255] She and many other Goddesses were named as "Lady of the Beasts." Swimme and Berry describe wildness as "a face ultimacy wears," "a primal act within the life process."[256] They say:

254 Swimme and Berry, *The Universe Story,* 125.

255 Spretnak, *Lost Goddesses of Early Greece*, 75.

256 Swimme and Berry, *The Universe Story,* 125.

> A wild animal, ... alert and free, moves with a beauty ... far beyond the lock-step process of a rationally derived conclusion. The wild is a great beauty that seethes with intelligence, that is ever surprising and refreshing ... The discovery of mutations is the discovery of an untamed and untameable energy at the organic centre of life. ... For without this wild energy, life's journey would have ended long ago.[257]

At Beltaine the Virgin's passion moves Her more deeply into engagement with the Other – Her face is noticeably changing into the Mother. Her desire for complete fullness continues to wax. Her movement, Her Lust, is to open completely into the Omega of Summer Solstice.

Summer Solstice/Litha – December 20-23 (S.H.), June 20-23 (N.H.)

The "moment of grace"[258] that is Summer Solstice, marks the stillpoint in the height of Summer, when Earth's tilt and orbit cause the Sun's energy to begin its decline – its movement back to the North in the Southern Hemisphere, and to the South in the Northern Hemisphere. This Seasonal Moment is polar opposite Winter Solstice when it is light that is "born." At the peak of Summer, in the radiance of expansion, it is the dark that is "born."[259] It is a celebration of profound mystical significance, which in a culture where the dark is not valued for its creative telios, may be confronting.

The purpose for the seasonal gathering is stated thus, in my own adaptation of Starhawk's version:[260]

---

[257] Ibid., 127.

[258] Thomas Berry's term for moments of transition: dawn and dusk, and seasons.

[259] In Australia and in other global locations the heat usually increases after Summer Solstice, so inhabitants are often wanting to close out the heat and light, and enter the relief of the Dark – which is resonate with the Solstice turning.

[260] Starhawk, *The Spiral Dance*, 205-206, with additions from 236 where she describes Summer Solstice as "the Give-Away time of the Sun."

> This is the time of the Summer Solstice in our Hemisphere – when the light part of the day is longest. In our part of the world, light is in Her fullness, She spreads Her radiance, Her fruits ripen, Her greenery is peaking, the cicadas sing. Yet as Light reaches Her peak, our closest contact with the Sun, She opens completely, and the seed of darkness is born.
>
> As it says in the tradition, this is the season of the rose, blossom and thorn, fragrance and blood. The story of Old tells that on this day Beloved and Lover embrace, in a love so complete, that all dissolves, into the single Song of ecstasy that moves the worlds. Our bliss, fully matured, given over, feeds the Universe and turns the wheel. We join the Beloved and Lover, Earth, Moon and Sun, in the Great Give-Away of our Creativity, our Fullness of Being.

Summer is a time for celebrating our realized Creativity, whose birth we celebrated at Winter, whose tenderness we dedicated ourselves to at Imbolc/Early Spring, whose certain presence and power we rejoiced in at Spring Equinox/Eostar, whose fertile passion we danced with at Beltaine/High Spring. Now, at this seasonal point, as we celebrate Light's fullness, we celebrate our own ripening – like that of the wheat/grain, and the fruit. And like the wheat/grain and the fruit, it is the Sun that is in us, that has ripened: the Sun is the Source of our every thought and action. The analogy is complete in that our everyday Creativity and we ourselves are ultimately also Food for the Universe.[261] Like the Sun and the grain and the fruit, we find the purpose of our Creativity in the releasing of it, in its consumption; just as our breath must be released for its purpose of Life. The ceremonial symbolism used to express this is the giving of a full rose or flower of choice to the flames,[262] or to a body of water. We, and our everyday Creativity are given over. In this way we each are the Bread of Life; just as many other

---

[261] This is a metaphor I learnt from Brian Swimme, *Canticle to the Cosmos*, video 5, and something that is understood in Indigenous traditions throughout the world.

[262] This also resonates with Summer being the season of fires, and something can be made of that in the ceremony.

indigenous traditions recognize everyday acts as evoking "the ongoing creation of the cosmos,"[263] so in this tradition, Summer is the time for particularly celebrating that. Our everyday lives, moment to moment, are built on the fabric of the work/creativity of the ancestors and ancient creatures that went before us. So the future is built on ours. We celebrate the blossoming of Creativity then, and the bliss of it, as Earth does pour forth Her abundance, gives it away freely. We aspire to follow Her example. In this cosmology, what is given is the self fully realized and celebrated, not a self that is abnegated – just as the fruit gives its full self: as Starhawk says, "Oneness is attained not through losing the self, but through realizing it fully."[264]

Summer Solstice is a celebration of the Fullness of the Mother – in ourselves, in Earth, in the Cosmos. It is the ripening of Her manifestation, which fulfills itself in the awesome act of dissolution. It is a celebration of Communion, the Feast of Life – which is for the enjoying, not for the holding onto. I represent this Seasonal Moment on my Wheel of the Year with a horseshoe, because its yonic shape may represent Goddess' "Great Gate",[265] Her gateway, as many cultures understood it, some building sacred sites in that shape. It is also the shape of the last letter of the Greek alphabet, Omega, literally 'great O', a great ending, a consummation. Summer Solstice is such a Gateway. At this interchange, the Virgin's face has passed through the Mother into the Crone. The process of the three Seasonal Moments of Beltaine, Summer Solstice and Lammas, as a group, may be felt as the three faces of Cosmogenesis in the movement towards entropy.

Lammas/Late Summer – February 1st (S.H.), August 1st (N.H.)

Lammas is the meridian point of the first dark quarter of the year, after the light phase is complete, and as such, it is a special celebration of the Crone/Old One. Within the Celtic tradition and for many Pagans, it is the wake of Lugh, the Grain God, the Sun King – the moment of marking his passing, the first harvest, named commonly as "Lughnasad"; and it is the Crone that reaps him. Within earlier Goddess traditions, all

---

263 Spretnak, *States of Grace,* 95.

264 Starhawk, *The Spiral Dance*, 37.

265 Walker, *The Woman's Encyclopedia of Myths and Secrets,* 414.

the transformations were Hers – which is also to say that all the transformations are within each being: that is, the "sacrifice," the passing, is not carried out by another external to the self, as has been commonly told and interpreted from stories of Lugh (or Jesus). In earliest gynocentric traditions,

> "the community reflected on the reality that the Mother aspect of the Goddess, having come to fruition, from Lammas on would enter the Earth and slowly become transformed into the Old Woman-Hecate-Cailleach aspect …"[266]

I dedicate Lammas to the quality/phase of the Crone/Old One; just as Imbolc, its polar opposite on the Wheel, is dedicated to the Virgin/Young One quality/phase. The Old One, the Dark and Shining One, has been much maligned, so to celebrate Her can be more of a challenge in our present cultural context. Lammas may be an opportunity to re-aquaint ourselves with the Crone in her purity, to fall in love with Her again.

I state the purpose of the seasonal gathering thus:

> This is the season of the waxing dark. The seed of darkness that was born at the Summer Solstice now grows … the dark part of the days grows visibly longer. Earth's tilt is taking us back away from the Sun. This is the time when we celebrate dissolution, the time when each unique beautiful self lets go, to the Darkness. It is the time for celebrating endings, when the grain, the fruit, is

[266] Durdin-Robertson, *The Year of the Goddess*, 143, quoting McLean, *Fire Festivals*, 20-22. And there are many other references for the earlier Goddess traditions beneath "Lughnasad": another is Gray, *The Woman's Book of Runes*, 18. The earlier tradition beneath "Lughnasad" is also indicated by the other name for it in Ireland of "Tailltean Games." Taillte was said to be Lugh's foster-mother, and it was her death that was being commemorated (Mike Nichols, "The First Harvest," *Pagan Alliance Newsletter*).

> harvested. We meet to remember the Dark Sentience, the All-Nourishing Abyss, She – from whom we arise, in whom we are immersed and to whom, we return.
>
> This is the season of the Crone, the Wise Dark One, who accepts and receives our harvest, who grinds the grain, who dismantles what has gone before. She is Hecate, Lilith, Medusa, Kali, Ereshkigal, Chamunda, Coatlique – Divine Compassionate One, She Who Creates the Space to Be. We meet to accept Her transformative embrace, trusting Her knowing, which is beyond all knowledge.

Lammas is the seasonal moment for recognizing that we dissolve into the 'night' of the Larger Organism that we are part of – "Gaia" as She may be named. It is She, the primordial Mother, however we name Her, who is immortal, from whom we arise, and into whom we dissolve. This celebration is a development of the dark that was born in the transition of Summer Solstice; the dark sentient Source of Creativity is honoured. The *autopoietic* space in us recognizes Her, is comforted by Her, desires Her self-transcendence and self-dissolution; Lammas is an opportunity to be with our organism's love of Larger Self – this Native Place. We have been taught to fear Her, but at this Seasonal Moment, we may remember that She is the Compassionate One, deeply committed to transformation, which is actually innate to us.

Whereas at Imbolc, we shone forth as individual, multiforms of Her; at Lammas, we small individual selves remember that we are She and dissolve back into Her. We are the Promise of Life as we affirmed at Imbolc, but we are the Promise of *Her* – it is not ours to hold. We become the Harvest at Lammas; our individual harvest *is* Her Harvest. We are the process itself – we are Gaia's Process. *We* do not breathe (though of course we do), we borrow the breath, for a while. It is like a relay: we pick the breath up, create what we do during our time with it, and pass it on. The harvest we reap in our individual lives is important, *and* it is for us only short term; it belongs to the Cosmos in the long term. Lammas is a time for "making sacred" – as "sacrifice" may be understood; we may "make sacred" ourselves. As Imbolc was a time for dedication, so is Lammas. This is the Wisdom of the phase of the Old One. She is the aspect that finds the "yes" to letting go, to loving the

Larger Self, beyond all knowledge, and steps into the *Power of the Abyss*, encouraged and nourished by the harvest – so She moves into the balance of Mabon/Autumn Equinox.

Autumn Equinox/Mabon – March 20-23 (S.H.), September 20-23 (N.H.)

Autumn Equinox, or "Mabon" as it may be named, is a time of thanksgiving for the harvest – for the empowerment and nourishment of the 'harvest' in its myriad forms; and it is also a time of leavetaking and sorrow, as life declines:[267] it is a balance of grief and joy. As the dark continues to grow, it comes into balance with the light. Autumn Equinox/Mabon is that point of balance in the dark part of the cycle: Sun is equidistant between South and North, as it was at Spring Equinox, but in this phase of the cycle, the trend is toward increasing dark. For millennia in Greece, this Seasonal Moment has been the holy celebration of the Daughter Persephone's descent to the Underworld; and as mentioned earlier, in the earliest Goddess tradition and in this PaGaian Cosmology, Her descent is voluntary – She simply understands the necessity of the Journey.

As I say in the statement of purpose for the seasonal ceremony,

> Feel the balance in this moment – Earth as She is poised in relationship with the Sun. See Her there in your mind's eye. Contemplate this balance. Feel for your own balance of light and dark within. Breathe into it. Breathe in the light, swell with it, let your breath go into the dark, rest with it. Feel for your centre... breathe into it – this Sacred Balance, from which all comes forth.
>
> In our part of Earth, the balance is tipping into the dark. Feel the shift within you, see in your mind's eye the descent ahead, the darkness growing, remember the coolness of it. This is the time when we give thanks for our harvests – the abundance we have reaped. And we remember too the losses involved. The story of Old tells us that Persephone, Beloved Daughter, receives from Her Mother, the wheat – the Mystery, knowledge of life and death –

---

267 Starhawk, *The Spiral Dance*, 208.

for this she gives thanks. She receives it graciously. But she sets forth into the darkness – both Mother and Daughter grieve that it is so.

Like its Spring counterpoint, Autumn Equinox is also a "stepping into power," but it is not necessarily perceived as such; it is a power felt as loss. Autumn Equinox is a time for grieving our many losses, as individuals, as a culture, as Earth-Gaia, as Universe-Gaia. At this time we may join Demeter, and any Mother Goddess from around the globe, in Her weeping for all that has been lost. The Mother weeps and rages, the Daughter leaves courageously, the Old One beckons with Her wisdom and promise of transformation; yet all three know Each Other deeply, and share the unfathomable grief. But Persephone as Seed represents the thread of life that never fades away. The revelation of the Seed, central to this seasonal celebration, is that:

> *Everything lost is found again,*
> *In a new form, In a new way…*
> *She changes everything She touches, and*
> *everything She touches changes.*[268]

And so it will be. In this way Persephone as Seed, tends the sorrows, 'wholes' the heart.

All at once, the three faces of Cosmogenesis are present. As Seed, She is Sovereign/Queen of the Underworld, Old Wise One, and also at once the irrepressible Promise/Urge to Be – they are never separate, the 'lost' Young One; and She is the Mother, Source of Life. This is a blessed Moment of harmony/balance that streams through the grief and the ecstasy of life.

This point of balance tips further into the dark, as Earth's tilt and Sun continue their relational dynamic. The dark of night keeps growing, the seed is in the Earth, the grub is in the chrysalis, the Abyss is accepted – the metamorphosis of the dark Sentience is awaited, the fertile emptiness of the Crone is the moment of Samhain.

With Samhain, the annual cycle – the Wheel of the Year – is

---

268 Ibid., 115.

complete. It is the time of death and re-conception, and the beginning of the New Year.

In Summary – Contemplating How Creativity Proceeds

In the flow of the PaGaian Wheel of the Year, the Seasonal transitions of the Wheel and the Triplicity of the Cosmos come together. There are two celebrations of the Old One/Crone or the Cosmogenetic quality of autopoiesis creating the *Space to Be*; and they are Lammas/Late Summer and Samhain/Deep Autumn, which are the meridian points of the two quarters of the waxing dark phase. At Lammas, the first in the dark phase, we may *identify* with the dark and ancient Wise One – dissolve into Her; at Samhain, we may consciously participate in Her *process* of the transformation of death/the passing of all. The whole dark part of the cycle is about dissolving/dying/letting go of being – becoming – nurturing it (the midwifing of Lammas/Late Summer), stepping into the power of it (the certain departure of Autumn Equinox/Mabon), the fertility (of Samhain/Deep Autumn), the peaking of it (at Winter Solstice).

The meridian points of the two quarters of the waxing light phase then are celebrations of the Young One/Virgin or the Cosmogenetic quality of differentiation, the new continually emerging, the *Urge to Be*; and they are Imbolc/Early Spring and Beltaine/High Spring. At Imbolc, the first in the light phase, we may *identify* with She who is shining and new – as we take her form; at Beltaine, we may consciously participate in Her *process* of the dance of life. The whole light part of the cycle is about coming into being: nurturing it (the midwifing of Imbolc/Early Spring), stepping into the power of it (the certain return of Spring Equinox/Eostar), the fertility (of Beltaine/High Spring), the peaking of it (at Summer Solstice).

In this PaGaian wheel of ceremony there are two particular celebrations of the Mother/Communion quality – the Solstices. If one imagines the light part of the cycle as a celebration of the Productions of Time, and the dark part of the cycle as a celebration of Eternity, the Solstices then are meeting points, and are celebrations of the communion/relational field of Eternity with the Productions of Time. This is a relationship which *does* happen in this Place, in this Web. This *Place of Being*, this Web, is a Communion – it *is* the Mother; the Solstices mark Her Birthings, Her Gateways.

The Equinoxes then – both Spring and Autumn – are two celebrations wherein the balance of all three Faces/Creative qualities is particularly present: in the PaGaian wheel, the Equinoxes have been special celebrations of Demeter and Persephone – echoing the ancient tradition of Mother-Daughter Mysteries that celebrate the awesomeness of the continuity of life, its creative tension/balance. Both Equinoxes then are celebrations and contemplations of empowerment through deep Wisdom – one contemplation during the dark phase and one during the light phase. The Autumn Equinox is a descent to Wisdom, the Spring Equinox is an emergence with Wisdom gained. I like to think of the Equinoxes, and of the ancient icons of Demeter and Persephone, as celebrations of the delicate "curvature of space-time," the fertile balance of tensions which enables it all.

The Mother aspect then may be understood to be particularly present at four of the Seasonal Moments, which are also regarded traditionally as the Solar festivals; and in this cosmology Sun is felt as Mother. I recognize these four as points of interchange: at Autumn Equinox, Mother is present primarily as Giver – She is letting Persephone go, at Spring Equinox, She is present primarily as Receiver – welcoming the Daughter back, at Winter Solstice the Mother gives birth, creates form, at Summer Solstice, She opens again full of radiance, and disperses form. The Mother is Agent/Actor at the Solstices. She is Participant/Witness at the Equinoxes, where it is then really Persephone who is Agent/Actor, embodying an inseparable Young One and Old One. The Old One is often named as Hecate, who completes the Trio – all seamlessly within each other.

Another possible way to visual it, or to tell the story, is this: The Mother – Demeter – is always there, at the Centre if you like. Persephone cycles around. She is the Daughter who returns in the Spring as flower, who will become fruit/grain of the Summer, who at Lammas assents to the dissolution – the consumption. At Autumn Equinox She returns to the underworld as seed – Her harvest is rejoiced in, Her loss is grieved, as She becomes Sovereign of the Underworld – Her face changes to the Dark One, Crone (Hecate). As the wheel turns into the light part of the cycle She becomes Young One/Virgin again. Persephone (as Seed) is that part of Demeter that can be all three aspects

– can move through the complete cycle. The Mother and Daughter/Child are really One, and embody the immortal process of creation and destruction. Demeter hands Persephone the wheat, the Mystery, and the thread of life is unbroken – it goes on forever. It is immortal, it is eternal. Even though it is true that all will be lost, and all is lost – Being always arises again: within this field of time there is never-ending renewal, eternity. This is what is revealed in the ubiquitous three faces of the Creative Dynamic/ She of Old, the Triplicity that runs through the Cosmos. The Seed of Life never fades away, She is always present; even though it may not be apparent. As Swimme and Berry, and other cosmologists note in their telling of story of our Universe, galactic clouds may drift for eons before undergoing transformation, but the fertility/potency seethes there.

As one participates in this year long process of ceremony the complexity of the three aspects does become more apparent. Indeed all three aspects/faces/energies do occur simultaneously – sometimes obviously so, as mentioned, sometimes perceived only at deeper levels; and at deeper levels, one can perceive how the three are features of each other. Usually, it is fairly clear that the Virgin and Crone aspects are inseparable, and this is true also of the Cosmogenetic characteristics of differentiation and autopoiesis. For example, differentiation/ diversity/ uniqueness has been enhanced by the advent of death in the biological story;[269] and the autopoietic dynamic is ambivalently self and not-self, it is small self who is larger Self at the same time. This inseparability of the new emergent form and the old sentient trans-form is congruent also with the ancient perception that a day begins in the evening before – as the dark space closes round. The severance comes first – is it birth or is it death? It is not always clear where something ends and where it begins.

There is a symmetry or reciprocity between the polar opposites on the Wheel, some of which I have already noted. One begins to feel these as one's familiarity with the Metaphor and the celebrations grows. Another relationship I have noticed is the continuity of Samhain-Winter

---

[269] See Sahtouris, *Earthdance,* 134-135 and Ursula Goodenough, *The Sacred Depths of Nature* (New York and Oxford: Oxford University Press, 1998), 143-149.

Solstice-Imbolc as the movement through the three faces towards form/syntropy, from Dark One to Birth to Light; and the continuity of Beltaine-Summer Solstice-Lammas as the movement through the three faces towards entropy/dis-order, from Light/Passion of Being to Radiance/Dispersal to Dark. This is a useful contemplation as it teaches how Creativity proceeds. When it is practised as a whole, it becomes a Gestalt, one may begin to Know: Her lap is a seat of learning – a throne.

CHAPTER 3
PAGAIAN CEREMONY

The Cosmos *is* a ceremony, a ritual. Dawn and dusk, seasons, supernovas – it is an ongoing *Event* of coming into being and passing away. The Cosmos is always in flux, and we exist as participants in this great ritual event, this "cosmic ceremony of seasonal and diurnal rhythms" which frame "epochal dramas of becoming," as Charlene Spretnak describes it.[270] Swimme and Berry describe the universe as a dramatic reality, a Great Conversation of announcement and response.[271] Ritual/ceremony[272] may be the human conscious response to the announcements of the Universe – an act of conscious participation. Ceremony then may be understood as a microcosmos[273] – a human-sized replication of the Drama, the Dynamic we find ourselves in. Swimme and Berry describe ritual as an ancient response humans have to the awesome experience of witnessing the coming to be and the passing away of things; they say that a "ritual mode of expression" is from its beginning "the manner in which humans respond to the universe, just as birds respond by flying or as fish respond by swimming."[274] It is the way in which we as humans, as a species, may respond to this awesome experience of being and becoming, how we may hold the beauty and the terror.

---

[270] Spretnak, *States of Grace*, 145.

[271] Swimme and Berry, *The Universe Story,* 153.

[272] I will use either or both of these terms at different times: I generally prefer "ceremony" as Kathy Jones defines it in *Priestess of Avalon, Priestess of the Goddess*, 319. She says that ritual involves a repeated set of actions which may contain spiritual or "mundane" elements (such as a daily ritual of brushing one's teeth), "whereas ceremony is always a spiritual practice and may or may not include ritual elements." The PaGaian seasonal celebrations/events are thus most kin to "ceremony," although I do not perceive any action as "mundane." However, "ritual" is more commonly used to speak of how humans have conversed with cosmos/Earth.

[273] Spretnak, *States of Grace*, 145.

[274] Swimme and Berry, *The Universe Story,* 152-153.

Humans have exhibited this tendency to ritualize since the earliest times of our unfolding: evidence so far reveals burial sites dating back one hundred thousand years, as mentioned in the previous chapter. We often went to huge effort in these matters, that is almost incomprehensible to the modern industrialised econocentric mind: the precise placing of huge stones in circles such as found at Stonehenge and the creation of complex sites such as Silbury Hill may be expressions of some priority, indicating that econocentric thinking – such as tool making, finding shelter and food, was not enough or not separate from the participation in Cosmic events. Ritual seems to have expressed, and still does actively express for some peoples, something essential to the human – a way of being integral with our Cosmic Place, which was not perceived as separate from material sustenance, the Source of existence: thus it was a way perhaps of *sensing* "meaning" as it might be termed these days – or "relationship."

Swimme and Berry note that the order of the Universe has been experienced especially in the seasonal sequence of dissolution and renewal; this most basic pattern has been an ultimate referent for existence.[275] The seasonal pattern contains within it the most basic dynamics of the Cosmos – desire, fullfilment, loss, transformation, creation, growth, and more. The annual ceremonial celebration of the seasonal wheel – the Earth-Sun sacred site within which we tour – can be a pathway to the Centre of these dynamics, a way of making sense of the pattern, a way of *sensing* it. One enters the Universe's story. The Seasonal Moments when marked and celebrated in the art form of ceremony may be *sens-ible* 'gateways' through the flesh of the world[276] to the Centre – which is omnipresent Creativity.

Humans do ritual everyday – we really can't help ourselves. It is simply a question of what rituals we do, what story we are telling

---

[275] Ibid.

[276] Abram speaks of "matter as flesh" in *The Spell of the Sensuous,* 66, citing Maurice Merleau-Ponty, *The Invisible and the Invisible* (Evanston, Illinois: Northwestern University Press, 1968).

ourselves, what we are "spelling"[277] ourselves with – individually and collectively.

Ceremony as "Prayer" or Sacred Awareness

Ritual/ceremony is often described as "sacred space." I understand that to mean "*awareness* of the space as sacred": all space is sacred, what shifts is our awareness – awareness of the depth of spacetime, and of the depth of all things and all beings. I understand "sacred awareness" as an awareness of deep relationship and identity with the very cosmic dynamics that create and sustain the Universe; or an awareness of what is involved in the depth of each moment, each thing, each being. Ceremony is a space and time given to expression, contemplation and nurturance of that depth … at least to *something* of it. Ceremony may be both an *expression* of deep inner truths – perceived relationship to self, Earth and Cosmos, as well as being a *mode of teaching* and drawing forth deeper participation.

Essentially, ceremony is a way of entering into the depth of the present moment … what is deeply present right here and now, a way of entering deep space and deep time, which is not somewhere else but is right here. Every-thing, and every moment, has Depth – more depth than we usually allow ourselves to contemplate, let alone comprehend. This book, this paper, this ink, the chair, the floor – each has a history and connections that go back, all the way back to Origins. This moment you experience now, in its particular configuration, place, people present, subtle feelings, thoughts, and propensity towards certain directions or outcomes, has a depth – many histories and choices that go back … ultimately all the way back to the beginning. Great Origin is present at every point of space and time – right here. In ceremony we are plugging our awareness into something of that.

In this holy context then – in this mindframe of knowing connection, everything one does is a participation in the creation of the Cosmos: for the tribal indigenous woman, perhaps the weaving of a basket; for another, perhaps preparing a meal; for you, perhaps getting on the train to go to a workplace. It is possible to regain this sense, to

---

[277] Starhawk used this term on her email list in 2004 to describe the story-telling we might do to bring forth the changes we desire.

come to feel that the way one breathes makes a difference – that with it, you co-create the present and the future, and you may even be a blessing on the past. In every moment we receive the co-creation, the work, of innumerable beings, of innumerable moments, and innumerable interactions of the elements, in everything we touch … and so are we touched by them. The local is our touchstone to the Cosmos – it is not separate. Ceremony may be a way into this awareness, into strengthening it.

Starhawk says that "to do ritual, you must be willing to be transformed in some way,"[278] because that is its very nature: that is, it is 'trans-forming'. Ritual can be as simple as having a cup of tea or reading a poem, or high drama like classical theatre or a rave concert; in either case it is "time out" – entering another realm, to a greater or lesser degree. As with having a cup of tea, it is done with the expectation of rejuvenation/renewal. Humans actually do ritual all the time. Starhawk notes that "ritual is the way culture enacts and affirms its values."[279] But this enactment and affirmation is usually unconscious, and thus the participants remain unaware of what is actually being activated; for example, going to the pub or bar is a popular sanctioned ritual – time out, imbibing 'spirit'. And whereas once, the ancestors used to sit around the fireplace and tell the cosmic stories; now most often humans sit around a television in the modern cave, and the story that is frequently told is that the world is a collection of objects to be consumed.[280] As ritual/ceremony is done consciously more often, we become aware of the symbols and myths that we live and can choose more consciously the tools with which we shape our lives. Ritual becomes ceremony; that is, *conscious*, and at its best it is the art form of a living cosmology.

Ceremony is actually 'doing,' not just theorizing. We can talk *about* our personal and cultural disconnection endlessly, but we need to *actually* change our minds. Ceremony can be an enabling practice – a catalyst/practice for personal and cultural change. It is not just talking *about* eating the pear, it is *eating* the pear; it is not just talking *about* sitting on the cushion (meditating), it is *sitting* on the cushion. It is a cultural

---

278 Starhawk, *Truth or Dare,* 100.

279 Ibid., 98.

280 See Swimme, *The Hidden Heart of the Cosmos*, 8-20.

practice wherein we tell a story/stories about what we believe to be so most deeply, about who and what we are. Ceremony can be a place for practicing a new language, a new way of speaking, or *spelling* – a place for practicing "matristic storytelling"[281] if you like: that is, for telling stories of the Mother, of Earth and Cosmos as if She were alive and sentient. We can "play like we know it," so that we may come to know it.[282] Ceremony then is a form of social action.

I have found it useful to describe ceremony using and extending words used by Ken Wilber to describe a "transpersonal practice," which is needed for real change: he said it was a practice that discloses "a deeper self (I or Buddha) in a deeper community (We or Sangha) expressing a deeper truth (It or Dharma)."[283] My extension of that is: ceremony may disclose a deeper beautiful self (the I/Virgin/Urge to Be/Buddha), in a deeper relational community (the We/Mother/Place of Being/Sangha), expressing a deeper transformative truth (the It/Old One/Space to Be/Dharma). This is the "unitive body," the "microcosmos" that Charlene Spretnak refers to.

According to Brian Swimme, the Universe is one huge celebration – expanding, exuberantly rushing away from a center with news of that center ... an urgent unfolding of being.[284] Thus, he says: "Self-expression is the primary sacrament of the universe. Whatever you feel deeply demands to be given form and released."[285] He describes this innate dynamic of celebration as a "generosity of being" that "insists upon song and dance." Ceremony must be a space where something deep in the self is free to be expressed – a space free of judgement and coercion – a space felt to be 'safe' which allows and invites individual uniqueness, while affirming community.

---

[281] A term used by Gloria Feman Orenstein in *The Reflowering of the Goddess* (New York: Pergamon Press, 1990), 147.
[282] As my doctoral thesis supervisor Dr. Susan Murphy once described it to me in conversation
[283] Ken Wilber, *A Brief History of Everything* (Massachusetts: Shambhala, 1996), 306-307.
[284] Swimme, *The Universe is a Green Dragon*, 144-145.
[285] Ibid., 147.

Since ceremony is an opportunity to give voice to deeper places in ourselves, forms of communication are used that the dreamer, the emotional, the body, can comprehend, such as music, drama, simulation, dance, chanting, singing.[286] These forms enable the entering of a level of consciousness that is there all the time, but that is not usually expressed or acknowledged. We enter a realm that is 'out of time,' which is commonly said to be not the "real" world, but it is more organic/ indigenous to all being and at least as real as the tick-tock world. It is a place "between the worlds," wherein we may put our hands on the very core of our lives, touch whatever it is that we feel our existence is about, and thus touch the possibility of re-creating and renewing ourselves.

The Place/Habitat

The site of the seasonal ceremonial celebrations will always be significant. In my case, the place in which I have created them has been notably in the Southern Hemisphere of Earth. The fact of my context being thus – the Southern Hemisphere – has contributed in the past to my deep internalized sense of being "Other," yet hence in the present this context contributes to my deep awareness of Gaia's Northern Hemisphere and Her reciprocal Seasonal Moment – the whole Planet. My initial confusion about the *sensed* Cosmos – as a Place, has become a clarity about the *actual* Cosmos – which remains inclusive of my sensed Cosmos. PaGaian reality – the reality of our Gaian "country" – is that the whole Creative Dynamic happens all the time, all at once. The 'Other,' the opposite, is always present – underneath and within the Moment. This has affected my comprehension of each Seasonal Moment, its particular beauty but also a fullness of its transitory nature. Many in the Northern Hemisphere – even today – have no idea that the Southern Hemisphere has a "different" lunar, diurnal, seasonal perspective, and because of this there often is a rigidity of frame of reference for place, language, metaphor and hence cosmology.[287] Indeed over the years of industrialized culture it has appeared to matter less to

---

286 As Starhawk notes, *The Spiral Dance*, 45.

287 Caitlin and John Matthews have been almost unique in their consideration of the Southern Hemisphere in their writing for decades. See *The Western Way*, 47.

many of both hemispheres, including then the 'author-ities,' the writers of culture and cosmos. Yet such 'author-ity' and northern-hemispheric rigidity has also been assumed by most more Earth-oriented writers as well: that is, there has been consistent failure to take into account a whole Earth perspective, and to acknowledge the relative nature of chosen metaphor. By way of example: the North Star does not need to be *the* point of sacred reference, as it has been classically in European Pagan traditions – there is great Poetry to be made of the void of the South Celestial Pole. Nor need the North direction be rigidly associated with the Earth element and darkness; nor is there really an "up" and a "down" cosmologically speaking. A *sense* and account of the Southern Hemisphere perspective with all that that implies metaphorically as well as *sens-ibly*, seems vitally important to comprehending and sensing a whole perspective and globe – a flexibility of mind, and coming to inhabit the real Cosmos, hence enabling a PaGaian cosmological perspective.

It has also been significant that my Search in its particular perspective, has been birthed in this ancient continent of Australia. It is the age of the exposed rock in this land, present to her inhabitants in an untarnished, primal mode that is significant. The land herself has for millennia been largely untouched by human war, conquest and concentrated human agriculture and disturbance. The inhabitants of this land dwelt here in a manner that was largely peaceful and harmonious, for tens of thousands of years. Therefore, I feel that the land Herself may speak more clearly; one may be the recipient of direct transmission of Earth in one of her most primordial modes, especially if one dwells in the country as I did, and as my grandmothers did. It seems to me that Earth's knowledge may be felt more clearly – one may be taught by Her. I think that the purity of this transmission, from my beginnings as a country girl – albeit below my conscious mind, in the subtle realms of which I knew little, to the more conscious times of entering into the process of the Search, is a significant factor in the development of the formal research I undertook – in my chosen methodology and in what I perceived in the process and documented. In this land that birthed me, "spirit" is not remote and abstract, it is felt in Her red earth.[288] Australian

[288] As Australian writer David Tacey also articulates in all of his work, but particularly referenced in "Spirit and Place," *EarthSong journal*, issue

First Nations elder David Mowaljarlai described: "This is a spirit country,"[289] and all of Her inhabitants, including non-indigenous, are affected by the strength of Her organic communication, when one hangs out with Her.
The specific region of the "Blue Mountains" as Europeans have named it (Gundungurra and Dharug country), is significant in that I don't think that this creative project could have happened as it did in just any region. David Abram says:

> The singular magic of a place is evident from what happens there, from what befalls oneself or others when in its vicinity. To tell of such events is implicitly to tell of the particular power of that site, and indeed to participate in its expressive potency.[290]

The Blue Mountains are impressive ancient rock formations, an uplifted ancient seabed, whose "range of rock types and topographical situations has given rise to distinct plant communities;"[291] and the presence of this great variation of plant communities, "especially the swamps, offer an abundance and variety of food sources, as well as habitats for varied fauna."[292] I feel that this was the case for this region's capacity to nurture this particular creative work as I have engaged in – it received the particularities of my passion. Even though I have been bringing a Western European heritage to this site, singing songs and dancing dances that come from other sites and times, it has been done in accord with the Seasons of this place and increasingly in accord with the particular features of this place. I learned recently that my specific local region in the Mountains is now thought to have been a ceremonial site for First Nations peoples.

My Search as it unfolded has been a journey of coming more deeply into relationship with my place, expressing and using the tools of

---

1, Spring 2004: 7-10 and 32-35: 7-9.

289 Quoted in Tacey, "Spirit and Place," 7.

290 Abram, *The Spell of the Sensuous*, 182.

291 Eugene Stockton, ed., *Blue Mountains Dreaming* (Winmalee: Three Sisters Productions, 1996), 43.

292 Ibid.

my ancestral heritage – knowing this heritage in myself first so that I may come into relationship. This Old European Pagan spiritual practice as I have re-created it here in Australia, is an imaginative response to the "challenge" of a "new cultural situation."[293] I have for some time felt 'familiar' with this place, related/family with this place, and in the course of my Search, this place has received and enabled, and indeed invoked the seed within me. I am aware that the Poetry of this work has been enacted and enabled within the specific context, and it would not have been the same elsewhere and in a different community, which is itself a creation of the place. And it will not be the same as it goes out, and others weave it into their own place and time.

Ceremony as a Poem

I understand any particular ceremonial event as an invitation to participate in a Poem, an invitation into a Poetic experience, much like going to theatre except that it is participatory theatre. It may be understood as a ceremonial theatre event wherein there is a script known in part by at least one facilitator/celebrant, and perhaps others, but it is not completely known, as the 'actors' are not yet all assembled with their full parts spliced in. There is an expressed intention in the gathering, but the actual unfolding is unpredictable and unknown. The Poetic experience is co-created though there is a guiding story – there is hearing and speaking, learning and expression, receiving and giving, that is ever reciprocal. The sum of the whole cycle of the year of ceremonial events then is an invitation to participate in a Poem over time – over the period of the year or years; and that has another whole dimension to it.

I have considered most of the specific phrases and language important, and it has not been a casual thing – I consider it Poetry; yet there is also some variation in the moment, as the particularity of the people, the person, the flow is sensed – and there is much variation in the responses, especially if the deep self has been encouraged and felt able to express. The scripts as I wrote them for each Seasonal Moment

---

[293] As David Tacey calls for, though it seems he has imagined that it would come mainly from old Christian traditions. See Tacey, "Spirit and Place," 9-10.

evolved and changed to some degree over time,[294] and continue to especially as others now co-create them. I am conscious of these scripts as having been a 'scribing' process, an 'authoring'; primarily as a Creative process that has been authored initially by me – though in-formed by others; and yet wherein each participant has been encouraged to articulate and 'scribe' their personal experience of this particular story. Generally, for many years only minor changes have been made to the scripts – though approach to some processes/rites have had many delightful variations; and the community of regular participants have expressed enjoyment of the consistency, the dependable 'pre-scription,'[295] finding that it allows a deepening into the Poetry.

Indigenous ceremonies – the ceremonies of traditional religious practices – aren't arbitrary: that is, they don't usually change much for perhaps thousands of years, in the story that is told within them, unless the purpose alters for some reason (which it has when climate changes, and/or peoples move, for example). The ceremonies remain somewhat predictable – pre-scribed: that is, certain truths about the moment are told, and the elders who have participated for a lifetime will generally know the depths and layers of these truths. So, it is for any seasonal ceremony that has elements repeated each year, the layers of the truths spoken may be understood/felt/known more deeply.

Artful Expression

The process of creating a beautiful space for ceremony, to paying artful attention to detail in its creation, and then also in the taking apart, pulling it down – both are each and together a process of learning. I like to think of it as a teaching gained "in the lap of the Mother": that is, the engagement in the crafting and the art of ceremonial preparation

---

[294] The process of the evolution of some of the ceremonial scripts is described in detail in Glenys Livingstone, *The Female Metaphor – Virgin, Mother, Crone – of the Dynamic Cosmological Unfolding: Her Embodiment in Seasonal Ritual as Catalyst for Personal and Cultural Change*, Chapter 7. *

[295] I do like to think of the seasonal ceremonies as a 'pre-scription' such as one might get from a doctor, for healing. It should be noted that 'pre-scribed' rituals or faiths may not all be healing: the script should be considered by each self.

will teach, and in the case of seasonal ceremonial creative preparation, it will teach the bodymind of the essential Creativity of this place we inhabit, as the Poetry embeds itself in action. And in the days or week following, seasonal decorations left around a home will nurture the deep truths spoken in ceremony, as one's eyes glance at them even in passing.

The changing and continuous seasonal decoration is a yearlong ceremonial art process of creation and erasure, kin to the construction of any other sacred artwork – for example the Kalachakra Mandala – that is created and destroyed. Any house decorations, any garments to be worn, any headpiece, a wreath, the altar, preparation of ceremonial foods, all participate in the expression and learning of the Seasonal Moment; then the decorations are removed, changed, for re-creation. Engagement in the art and craft process itself teaches – the art is ceremony and the ceremony is art. The *whole* ceremonial process – including the lengthy and detailed preparation, and the ceremonies themselves have been and are, central to the changing of bodymind – to the 'con-course' (it is more than a 'discourse').[296] The whole of these ceremonial processes, as the seasonal decoration changes over the year, are a major *Place of Conversation* – Cosmic Conversation,[297] that will indeed create a world – in the self, and hence perhaps in others and all. I have called this *Place of Conversation* the "Womb of Gaia"; as it is such a place of creation.

A special ceremonial headpiece that I decorated and wore for years became a personal central representation of the yearlong ceremonial art process of creating, destroying and re-creating. For a significant period, it came to represent for me the essence of 'She' – as Changing One, yet ever as Presence – as I was coming to know Her. In my journal for the Autumn Equinox/Mabon process notes one year I wrote:

---

296 "Con-course" is a term used by Jürgen Kremer in "Post-modern Shamanism and the Evolution of Consciousness," a paper delivered at the International Transpersonal Association Conference, Prague, June 20-25, 1992.

297 "Conversation" is a term used by Swimme and Berry to describe human ritual as responsive participation in the Universe, *The Universe Story*, 153.

> As I pace the circle with the Mabon headpiece in the centre, I see 'Her' as She has been through the Seasons … the black and gold of Samhain, the deep red, white and evergreen of Winter, the white and blue of Imbolc, the flowers of Eostar, the rainbow ribbons of Beltane, the roses of Summer, the seed pods and wheat of Lammas, and now the Autumn leaves. I see in my mind's eye, and feel, Her changes. I am learning … The Mother knowledge grows within me.

All the artful and crafted expression is part of the learning, the method, the relationship – similar to how one might bring flowers and gifts of significance to a loved one at special moments. Then further, the removal and re-creation of the decorations are part of the learning – an active witness to transformation through time.

Celebrant and/or Co-celebrants as Evocators of Presence

In presiding as Mistress of Ceremonies/Priestess in the ceremonies, I have been conscious of the importance of my voice, my invocation of the reality of which I speak, when I speak. Spoken words – everyday ones – are actually always an invocation, a spelling of reality, but in ceremony this may become more conscious. The celebrant and/or co-celebrants are participating in the calling forth of Presence within the consciousness of those present, calling forth the sacred depths. They will be enabled in that process by prior meditations upon, and awareness of, the truth of the Poetry they speak. They will also be enabled by the confident use of their breath and voice, which may involve some singing or voice lessons, and expressive therapies.

All others present participate in the calling forth of this Presence, in themselves and in others, and may take roles in this regard in some of the processes. It is my understanding that the Poetry itself has its own integrity, as does the speaker, and as do the recipients. The Poetry is not a monolithic inert slab of information; it speaks to our depths in a relational particularity – each person 'selecting' its valency for them. As Charlene Spretnak tells it in *States of Grace*, in the ritual creation of sacred space, the narrator and listener become "engaged witnesses, weavers of a web of being" where articulations of mythic dramas are "acts of relation that place all participants in deep accord with the life processes of the unfolding universe."[298] The role of priestess or celebrant is a deeply relational role; it is not just putting out inert questions, statements. She is evoking, drawing forth, and is already in response to the particular person's being – how they select/elicit a particular approach from her. The celebrant and/or co-celebrants listen deeply to the response, so the person is received. Sometimes, a particular invocation/blessing may be passed around the entire group, allowing individuals the opportunity to bless another, and to speak it. Thus, it is always more powerful when each participant has actually done some contemplation of the Poetry they speak: but the very speaking may be a potent beginning for anyone. This kind of participation may also be structured into the creation of the sacred space – the remembering of the elements of Water, Fire, Earth and Air – whenever possible, enabling individuals to practice speaking with or "*to* the world," instead of *about* it:[299] that is, to practice PaGaian relationship.

Ceremonial Format

I offer somewhat practical guides as to how I have been doing the ceremonial celebrations of the Seasonal Moments, as a means to embody the *Wholly Creative Dynamic* – to get with Gaia's "plot" as I see it. I understand that there is nothing necessarily pre-scriptive or "correct" about the way I have been doing the ceremonies, but I offer scripts as a possible way to proceed, and as an illustration of how one can and might put such ceremonial celebration together. One does not need to be a hierophant, a trained ordained person to begin: it may even be an advantage not to be,[300] although care is essential.

---

[298] Spretnak, *States of Grace*, 141.

[299] Abram, *The Spell of the Sensuous*, 71.

[300] This is not to denigrate or deny the wealth that comes from training and years of experience, it is simply to recognize that many ordained persons or qualified witches/priestesses may not necessarily facilitate as nurturant a sacred space as they are 'authored' to do. A person who has less 'hocus-pocus' (that is, "correct" terminology, holy robes and tools), but more care, real commitment to the Cosmos and simple psychological skill, *may* facilitate a more deeply affective ceremonial space.

Overall, it has been very important to me to make the ceremonial space one of 'ease'; that is, accessible and comprehensible to people, to allow it to be relevant to people's experience. Yet without compromising its sacred depths and solemnity, nor compromising the story that I wanted to spell out – which was also the story the participating people *wanted* to have spelled out. As much as possible, in any group preparation for the ceremonies, I have de-constructed the language often used in traditional Pagan circles to describe the ceremonial/ritual processes. My sense of freedom to do this is partly drawn from Starhawk's encouragement and style when I was taught by her. It is also partly drawn from an early background in de-constructing Catholic rituals (known as "Mass") almost a decade before that, when the people in that context began to take spiritual expression into their own hands. And so, it is now: the people in the larger community are increasingly desiring to take spiritual expression into their own hands. My priorities in ceremony are also affected by my training in liturgy at the Jesuit School of Theology in Berkeley, where we were taught carefully to enable the real Presence in the people.[301] These days I interpret that to mean enabling participants to bring their personal and collective stories to the story of the Seasonal Moment ceremony and therein to find deep communion with self, other and all-that-is.

The format/structure that I have used for the ceremonies is based on Starhawk's format for the traditional seasonal ceremonies that she outlines in *The Spiral Dance.*[302] Everyone or every group will have their own preferred emphases, additions, and varied understandings of the processes: a great deal of variety is possible, probably infinite. Briefly, the processes – the order of service – as I have done them for PaGaian ceremony flowed as below. Any of it can be adapted to an on-line format as may be needed/desired for ceremony in these times.

"Warming the Energy" – This is partly done in the preparation of the space before the event – with decoration appropriate to the Seasonal Moment, with a sense of presence conjured by aroma and music, and

---

301 This was the instruction and practice of James Empereur S.J. specifically.

302 Starhawk, *The Spiral Dance*, 197-213.

meditations that may be done there during the weeks or days before, as well as at the time of the event. The space is also "warmed" in the greetings and conversation that goes on as the gathering of participants happens, and as they select their positions in the prepared circle and place their ceremonial accessories (masks, stones, photos, flowers or whatever depending on the Seasonal Moment) in designated areas. There is eating and drinking of a light kind that goes on at this time also – in the kitchen, as part of the welcoming and settling in. The "warming" is also done in the group preparation for the ceremony, which is when I regard the circle as formally beginning, and the point at which the doors are closed to any latecomers. Participants are then smudged/smoked as they enter the ceremonial space, for the group preparation, which is essentially the time for going over the order of service, and learning the dances and/or songs, followed by about ten minutes of individual reflection /meditation before we re-gather for the actual ceremony; that ten minutes may also be for taking care of personal needs, so the ceremonial circle can proceed uninterrupted.

Call to Gather – This is the formal gathering moment: a formal call to gather to the circle, and it is a calling not just of the tangible biological selves, but also the mythic selves, and others in the subtle realms and all supportive energies, across space and time. It is usually done with a drum, which should continue at least until all are gathered in the circle, and perhaps a little longer, allowing time to feel the energy of the calling of all. During the call someone should be checking practical and safety details – like snuffing unattended candles and closing doors.

Centering or Breath Meditation – A breath meditation is the usual method of "Centering/Grounding" at the beginning of the ceremonies. It is a process of establishing connection with ourselves as 'breath-taking' beings, becoming present to the moment, moving into a deeper reflective

space. The way I have languaged them follows the flow of Gaia's breath over the year – connecting the individual breath cycle to the larger Creative cycle. I focus on the part of the breath pattern that matches the particular Seasonal Moment, sometimes adding quite extensive poetic flourish that introduces the theme of the ceremonial celebration: for example, for the Winter Solstice breath meditation there is reflection on the constant birthing in self, Earth and Cosmos – every day/moment throughout the eons and connecting that to the breath – "none of it separate, birthing you in every moment."

Statement of Purpose – stating the reason for the gathering, which in this case is the celebration of a Seasonal Moment; relating it specifically to both the sensed Earth-Sun experience, and the "story of Old." All the statements that I have written into the scripts are based in theme and expression on Starhawk's statements,[303] with my own additions and changes that include emphasis on Cosmogenetic and female metaphor. I am always careful to express what is actually happening with the light part and the dark part of the day, and to express it in this way; that is, with language that acknowledges the 'day' as diurnal – having a 'light part' and a 'dark part' that is in dynamic movement through the year. Such expression participates in spelling the dark back into the 'day'.

Creating/Recalling Sacred Space – traditionally this is often known as "the calling of the directions" and their associated elemental powers, and "casting the circle." The four elements of Water, Fire, Earth and Air are represented in the centre altar – posted in associated directions that may vary according to the Place of celebration.[304] The recognition of the directions signifies the presence of the Whole Cosmos here and within each one: we are each indeed recycled bits from everywhere and everytime. We don't have to go anywhere, it is all 'right-here' within each participant: we simply have to recall it. The essence of this ceremonial process then, as I see it, is the remembering of our common origins, remembering who and what we are, and from whence we come, thus creating awareness of a deeper space and time in which we all participate. It de-constructs our usual personas and social complexities, taking participants into a deeper more basic reality of being – which we share in common, and in which it is safe to speak deep truths and be heard and

303 Ibid.

304 This variation of direction for the associated elements is one of those realities particular to the mind of Southern Hemisphere people who have been transplanted within the last couple of hundred years from the Northern Hemisphere, but applicable to all locations.

known. This can be done with lots of poetry and drama, or with a simple group chant. As I have done them, there is almost always a direct involvement and identification of each participant with each element. Generally, individuals are invited to add in their own words to the addressing of the elements, in addition to the formal script which is following a Seasonal theme in its pattern.

Always I conclude this process of "casting the circle" with a statement like: "This is the Centre of Creativity," drawing attention to the fact that the Centre of the Universe is here,[305] that the *Ultimate Sentience* of the Universe is present. This is true of every place, being and time, and yet the ceremony is a space wherein this is made conscious, is what "casting the circle" means; that is, deeper awareness of being at Centre, in the ever present awesome Creative Space.

"Invocation" – after setting up the safe space in which deeper truths may be spoken, it is traditional to "invoke the divine" in whatever way that is understood. It may be considered a recalling that we are each expressions of something awesome, that in fact the truth of each person is that "here is the Source of everything, here is the Ultimate Mystery of the Universe."[306] This can be done in any number of ways ... poetry, anointing, a simple gesture shared in pairs or with the group. The method and approach also varies according to the Seasonal Moment, and the aspect of *Sacred Self* being addressed. It is an invocation of deep presence of each.

Quite often as celebrant I have spoken it to each person which may be of some benefit if a group is largely inexperienced. Just as often participants will pass it to each other, and there is benefit to that also, as each has the opportunity to practice the speaking. Generally, people can memorize any words that are used adequately, with some improvisation and prompting. Ideally, if whatever is said, is said with conviction, a deep knowing – it is an invocation of Presence. As I see it, that is what

---

[305] We live in an omnicentric Universe, noted by Swimme, *The Hidden Heart of the Cosmos,* 85-87 and 90.

[306] Swimme, *Canticle to the Cosmos*, video 1, "The Story of Our Time." Note that the stating of this to each participant may constitute the process of invocation.

"transmission" is: that is, if a person can tell another with a deep knowing that "Here is the Source of everything, here is the Ultimate mystery of the Universe," the receiver will most likely experience it on some level. That is what "blessing" is supposed to be, what invocation is. It is a deep knowing that the Power of the Universe is present. That, to my mind, is what it means to be "between the worlds," as traditional Pagan language describes the ritual/ceremonial space. This deep knowing is nurtured by meditation and practice over time, as well as by good teaching, and will usually be felt more deeply within a ceremonial context of experienced participants.

The Invocation as I have always done it, includes either a response from each participant that is central to the process, an affirmation that they accept this as a truth, or sometimes the Invocation is scripted as a space for each to announce themselves – in some way identifying themselves as a Larger Self, in accord with the Seasonal theme. Participants are encouraged to write or speak their own words, or at least to add their own words to any formal pre-scripted ones offered. I have found that it is also important in the preparation of the group for this process, to verbally offer participants the freedom not to speak, but to simply gesture in some way so that they may be welcomed as a Presence. Some may be far too shy; others may simply not have words – there may be any number of reasons for choosing a silent but conscious gesture.

Seasonal Rite or Body of the Ceremony – this is the particular dramatic processes that give expression to the Seasonal Moment. These are of great variation, according to personal, communal and regional aesthetics and factors. It includes what is traditionally known as "raising of energy" and the "working of magic"; that is, the transformational processes that are possible in ceremonial space. There are some possibilities offered in the ceremonial scripts for each Seasonal Moment in following chapters.

Sharing of Food or Communion – This is a formal recognition of how we are sustained, what gives us life, and of the fundamental communion experience of the Universe. It is a time within the ceremony for thanksgiving, some relaxation and enjoyment. The food that is shared varies, and it is given to each participant formally with an invocation/"blessing" – sometimes from the celebrant or celebrants, sometimes from each other as it is passed around the circle. The food and the process should connect with the theme of the Season.

Storytelling – a listening circle, wherein participants may speak to the group without having a discussion or being argued with. Any response from the group is formal and brief, such as "we hear you," "may it be so" or "thank you," or some other appropriate empathetic affirmation of having heard the speaker. It takes some discipline and sensitivity for participants to restrict themselves to hearing and only responding formally; usually the nature of the process has to be mentioned during the ceremonial preparation and re-iterated when introduced during the ceremony. Often the celebrant may need to keep drawing the group consciousness back to the formal agenda with further invitation to others who may wish to speak. Sometimes patience and tolerance are required as some participants seem to go on, and not all may find the story being told pertinent. Yet this space is always an exercise in respect, sensitivity, true receptivity, and group trust. Each seasonal ceremonial script offered in following chapters has a suggested agenda for this process that relates to the seasonal theme, but it is really an open space for whatever individuals need to speak of.

Opening of the Circle – a retracing of the "Sacred Space" invocations and associated elements, and a summarized recapitulation of the seasonal ceremonial process that has been participated in, with a peace blessing. Then the final words that I use are my version of the traditional ones[307]– sometimes sung, sometimes spoken:

> The circle is open but unbroken. May the peace of Goddess go in our hearts. It has been a merry meeting. It is a merry parting. May we merry meet again. Blessed Be.

Often a kiss may be passed (sometimes in both directions!), but this would depend on the nature of the group.

---

[307] See Starhawk, *The Spiral Dance*, 185.

Conscious Ceremonial Celebration of Cosmogenesis

Woven throughout the Seasonal ceremonial celebrations may be the conscious celebration of Cosmogenesis, the Creative unfolding of *Gaia-Universe&Earth*. These ceremonial celebrations begin with Earth-Sun relationship – that is the reason for their existence since the earliest of human times. The resulting Creativity of the play of light and dark in this Earth-Sun relationship has translated into food, and into human psyches. The creative telling of our personal and collective stories, and how we wish them to unfold, may be folded into the Seasonal Moment because that is where each participant may interface intimately with Gaia. Then also, as participants in the Larger Story of Gaia – *knowing* this is our full story – there is always this deeper layer to be expressed, and it may be drawn specifically into the Seasonal Moments. I will suggest some of the connection between each Seasonal Moment's theme and the Universe Story in the following chapters, and how that relationship may be expressed specifically in the ceremony.

On the surface of it, the dark Old One/Crone phase particularly celebrates the quality of autopoiesis – sentience, subjectivity, interiority, the creative centre; the light Young One/Virgin phase particularly celebrates the quality of differentiation – diversity, complexity, multiform nature, articulation. Communion – the Mother quality, reciprocity, deep relatedness, interconnectivity, mutuality – is celebrated somewhat consciously throughout (since She **is** this 'Sacred Interchange,' this Event), though particularly at the Solstices, and in balance with other two qualities at the Equinoxes. There may be many ways of folding in aspects of Gaia's story, of Cosmogenesis, expressing resonant moments of her Story that one wants to celebrate.

The Wheel of the Year as a "Turas"

The eight points of the wheel in this and many Pagan calendars represent 'seasonal thresholds' (and there may be more or less in your region), the circuit of Earth about the Sun. This is the sacred site in which we, all this planet's beings, find ourselves, in which we live every day. We may think of this journey around Sun as the revolving walk of a pilgrim about a sacred site – what the Celts called a "turas." This circle of eight stones/objects represents this sacred Journey. *Turas* is a Celtic word meaning "journey," "pilgrimage," and refers especially to the circular, spiralling prayer used by people in Celtic countries as they walked sunwise around a sacred site[308] (and sunwise in the Southern Hemisphere is counterclockwise).

The ceremonial celebration of the "Wheel of the Year" as it manifests in your place, as a whole year-long experience, participated in fully as an art process and relationship with Gaia, IS a sacred site – a kind of virtual sacred site, a morphic field: that is, the ceremonies themselves develop a kind of organism, an alive space (a womb). You can be held within it. One may enter consciously into this sacred site – real space and time – through the practice of ceremonial celebration of this annual journey, the *Turas* of our planet: and thus enter into the magic and power of deep Creativity – found in real time and space. For this reason I am religious about not doing festivals dictated by the Gregorian calendar, especially since they are out of sync with southern hemispheric seasons – Christmas, Easter. I am on a Journey, a pilgrim in real time and space, and indeed I have learned so much with this creative discipline … I am Her disciple.

Below is a walking meditation, that I have developed to create a mini-experience of this everyday turas. It may be done around a simpler version of a Womb of Gaia altar as pictured in the Introduction.

TURAS EXPERIENCE – a walking meditation

NOTE: if you are unable to walk around your altar that is okay … visualize the process as you sit. Begin sitting at the outer edge of your circle then move into the center as directed in the meditation, if you are able to.

We begin walking sunwise (counter-clockwise for Southern Hemisphere), starting some distance out from your altar if possible.

DRUM if you have one

Let us begin, by walking sunwise, slowly around the edge of your circle, gazing upon the eight points of the mandala … aware that you are passing through the Seasons, as you always have done, every day of your life, joining Earth in Her everyday sacred journey. You are Earth making this

---

308 Matthews, *The Celtic Spirit*, 31.

Journey.

After you have made a few circuits, begin to spiral in slowly and contemplating this particular Moment – all that it has taken for you to come to this Place and Moment … beginning with taking the time to be with this process, then all that led to your decision to do this, your personal story that brings you to this, the stories of your parents, of your grandparents … and further back … to stories of ancestors and other ancestral beings, who have walked this turas of Earth around the Sun – joining all these.

… and contemplating where your ancestors may have come from, and where every atom in your body may have been … and slowly spiralling into the Centre.

When you have arrived at Centre, you may consider this Centre, THE Centre, our Origin, which is always present.

Ten percent of your bodymind which is hydrogen, is a direct result of the Original Flaring Forth, when all hydrogen was made. The Origin is present right here within you. And all the rest – carbon, oxygen, nitrogen, phosphorus, and heavier elements – was born in stars. All this recycled many times over.[309]

Here in the Centre, you may celebrate Cosmogenesis, who you are
PLACE YOUR HANDS ON YOURSELF

- your particular beautiful Self, new in every moment – Virgin

OPEN UP YOUR ARMS

- in deep relationship and communion with Other, the web of life – Mother

[309] This paragraph is a quote from Australian molecular biologist Darryl Reanney when he gave an experiential paper at The Climbing River Foundation conference in Melbourne in 1990.

DIRECT YOUR HANDS AND ARMS DOWN TO EARTH

- directly participating in the sentience of the Creative Cosmos, the Well of Creativity – Old One.

FOLDING YOUR ARMS OVER YOUR TORSO
All present here … this Creative Dynamic unfolding the Cosmos.

And now, with this memory, turning to your left (right for the Northern Hemisphere) and slowly spiraling back out to the circumference, to your place in this Moment of time.

DRUM … slowly spiraling back out to the circumference, to your Place in this Moment of Time.
DRUM until you/all have arrived at the circumference.

And Here you are! And so, it is for every moment.

CHAPTER 4
SAMHAIN/DEEP AUTUMN

Northern Hemisphere – October 31st/November 1st
Southern Hemisphere – April 30th/May 1st
These dates are traditional, though the actual astronomical date varies. It is the meridian point or cross-quarter day between Autumn Equinox and Winter Solstice, thus actually a little later in early May for S.H., and early November for N.H., respectively.

In this cosmology, Deep Autumn/Samhain is a celebration of *She Who creates the Space to Be* par excellence. This aspect of the Creative Triplicity is associated with the *autopoietic* quality of Cosmogenesis[310] and with the Crone/Old One of the Triple Goddess, who is essentially creative in Her process. This Seasonal Moment celebrates the *process* of the Crone, the Ancient One … how we are formed by Her process, and in that sense conceived by Her: it is an 'imaginal fertility,' a fertility of the dark space, the sentient Cosmos. It mirrors the fertility and conception of Beltaine (which is happening in the opposite Hemisphere at the same time – see diagrams in Chapter 2).

Some Samhain/Deep Autumn Story

This celebration of Deep Autumn has been known in Christian times as "Halloween," since the church in the Northern Hemisphere adopted it as "All Hallow's eve" (31st October) or "All Saint's Day" (1st November). This "Deep Autumn" festival as it may be named in our times, was known in old Celtic times as Samhain (pronounced "sow-een), which is an Irish Gaelic word, with a likely meaning of "Summer's end," since it is the time of the ending of the Spring-Summer growth. Many leaves of last Summer are turning and falling at this time: it was thus felt as the end of the year, and hence the New Year. It was and is noted as the beginning of Winter. It was the traditional Season for bringing in the animals from the outdoor pastures in pastoral economies, and when many of them were slaughtered.

Earth's tilt is continuing to move the region away from the Sun at this time of year. This Seasonal Moment is the meridian point of the

---

310 Swimme and Berry, *The Universe Story*, 75-77.

darkest quarter of the year, between Autumn Equinox and Winter Solstice; the dark part of the day is longer than the light part of the day and is still on the increase. It is thus the dark space of the annual cycle wherein conception and dreaming up the new may occur. As with any New Year, between the old and the new, in that moment, all is possible. We may choose in that moment what to pass to the future, and what to relegate to compost. Samhain may be understood as the *Space* between the breaths. It is a generative Space – the Source of all. There is particular magic in being with this *Dark Space*. This Dark Space which is ever present, may be named as the "All-Nourishing Abyss,"[311] the "Ever-Present Origin."[312] It is a generative *Place*, and we may feel it particularly at this time of year, and call it to consciousness in ceremony.

The fermentation of all that has passed begins. This moment may mark the *Transformation of Death* – the breakdown of old forms, the ferment and rot of the compost, and thus the possibility of renewal.[313] It is actually a movement towards form and 're-solution' (as Beltaine – its opposite – begins a movement towards entropy and dissolution). With practice we begin to develop this vision: of the rot, the ferment, being a movement towards the renewal, to see the gold. And just so, does one begin to know the movement at Beltaine, towards expansion and thus falling apart, dissolution. In Triple Goddess poetics it may be expressed that the Crone's face here at Samhain begins to change to the Mother – as at Beltaine the Virgin's face begins to change to the Mother: the aspects are never alone and kaleidoscope into the other … it is an alive dynamic process, never static.

---

[311] Brian Swimme's term for the void out of which all arises, *The Hidden Heart of the Cosmos*, 97.
[312] Jean Gebser, *The Ever-Present Origin* (Athens Ohio: Ohio University Press, 1985).
[313] Reference may be made to the PaGaian Wheel of the Year diagram in Chapter Two, for Samhain's placement and context.

Samhain/Deep Autumn Breath Meditation

A good place to begin contemplation of this or any Seasonal Moment is with the breath:

> Centering for a moment … breathe deep, and as you let go of your breath, follow it down, and notice the space between your breaths. Don't hold it, just notice it. And again, breathing deep and following the breath down to this space – that is always with you, and feeling it. And again, breathe deep and as your breath empties, noticing this space … into which all who have gone before us have travelled … from which to enter again the dance of life in some form. Feel this space between your breaths – a sentient fertile Space.

Some Samhain/Deep Autumn Motifs

[Figure 9] Womb of Gaia altar with motifs added to the Samhain stone

Suggested decorations for this Seasonal Moment, placed on the front door, windowsills, corners of shelves, and/or altars, may include:
– gold fabric – representing transformation, which death is from Gaia's larger perspective.
– thin fabric – the veils are thin between the old and the new.

– snake – representing the shedding of the old, and life-force and regeneration.
– apple – representing 'never-ending renewal' (a more Earth-based term for 'ever-lasting life' or 'immortality'). The seed is within the fruit, and the fruit is within the seed.
– thread (natural fibre string) represents the spinning/weaving of the new, and it is meant to reflect the spider's fibre from her own body: essential creativity.
– thin gold ribbon – representing the 'gold' in the compost, if one has the vision for it.
– scissors – for cutting the thread, an ending and free.
These motifs are in included in the ceremonial script offered, and more detailed notes made about their significance.

An Invocation
SAMHAIN[314]

Changing now, changing quickly now,
going into the darkness
letting go of knowing.
Allowing myself to be subject
to the fertility of the dark.

Allowing a new pattern to emerge.

Letting the rich darkness soak through my being
like blood;
so a new constellation can gestate,
twinkle through to me
from an ancient past/future time,
from a galaxy within.

Allowing a new constellation
to form in the Void;
to guide my ship by.

---

[314] A poem by Glenys Livingstone, 1995.

As the New Year turns – the New Era turns,
letting go of the past
with all its learnings and gifts

thankfully and graciously.

Notes on Samhain/Deep Autumn Metaphor

The whole Wheel is a Creation story, and Samhain is the place of the **conceiving** of this Creativity, and it may be in the *Spelling* of it – *saying* what we **will**; and thus, beginning the Journey through the Wheel. As I express it my doctoral thesis:

> My primary vision of the Wheel's Pattern is one of Creative Power. The Female Metaphor, in Her triple aspects, is a metaphor of Creativity, based in actual life processes. The evolutionary cosmic dynamics – Cosmogenesis with its triple aspects – is an actual Creative process, a "physic" of the Universe. The Seasonal Moments of Gaia express this actual Creativity. The three together – the Female Metaphor, Cosmogenesis, and the celebration of the Seasonal Moments – are a Language of Creativity, a Poetry that can express and create.
>
> Language is a primary habitat of the human (Swimme 1990, video 9). Wisdom traditions have always understood the power of speech and image; the Western technological-scientific culture has forgotten a reverence for this capacity. The Wheel of the Year as it has been celebrated in my research has been a remembering of this power as a Power, a consciousness of what we are 'spelling' out. It has been a "spell of the sensuous"[315] – of the 'sens-ible,' wherein the language has been regarded as 'material' – as hard as rocks – as 'real'. The participants and I have spoken, danced and dreamed a Language, a Metaphor that we understood as resonant with the evolutionary cosmic dynamics. All speech participates in the creation of something; it may participate in the Creative Act, which is ongoing in every moment, at one with every breath, and action. I sense the celebration of the Wheel of the Year – the

---

[315] Abram's term, *The Spell of the Sensuous.*

> Seasonal Moments – as a Creative Gaian Power wherein our "small daily acts" and the way we breathe, makes a huge difference.[316]

That is the premise: where we begin.

This Seasonal Moment is a face of the Crone/Old One moving into Mother – a face of the Crone/Old One as Creator, how Her transformation of death is truly a Moment of *Creative Conception*. This is the autopoiesis of the Cosmos – its sentient capacity, the "Well of Creativity," where spontaneities seethe in the quantum foam. In traditional language as Starhawk has phrased it in her Samhain ceremonial statement of purpose, we journey "over the sunless sea that is the womb of the Mother" and step ashore "the luminous world egg," "the Shining Isle," and become the seed of our own rebirth.[317] The "Shining Isle"/"luminous world egg" may be understood as the ovum in the womb, where the new is conceived. Such a journey is played out in the offered ceremonial script in this chapter.

Conception

Conception could be described as a "female-referring transformatory power" – a term used by Melissa Raphael in *Thealogy and Embodiment*:[318] conception happens in a female body, yet it is a multivalent cosmic dynamic, that is, it happens in all being in a variety of forms. It is not bound to the female body, yet it occurs there in a particular and obvious way. Androcentric ideologies, philosophies and theologies have devalued the event and occurrence of conception in the female body: whereas PaGaian Cosmology is a conscious affirmation, invocation and celebration of "female sacrality"[319] as part of all sacrality. It does thus

---

316 Livingstone, *The Female Metaphor – Virgin, Mother, Crone – of the Dynamic Cosmological Unfolding: Her Embodiment in Seasonal Ritual as Catalyst for Personal and Cultural Change,* 333-334. Barbara Ardinger spoke of "the way we breathe, making a difference" in *A Woman's Book of Rituals and Celebrations* (New World Library, 1995).

317 Starhawk, *The Spiral Dance*, 210.

318 8-9.

319 Ibid., 8.

affirm the female as **a** place; as well as a **place.**[320] 'Conception' is identified as a Cosmic Dynamic essential to all being – not exclusive to the female, yet it is a female-based metaphor, one that patriarchal-based religions have either co-opted and attributed to a father-god (Zeus, Yahweh, Chenrezig – have all taken on being the 'mother'), or it has been left out of the equation altogether. Womb is the place of Creation – not some God's index finger as is imagined in Michelangelo's famous painting.

Melissa Raphael speaks of a "menstrual cosmology":

> As the Celtic goddess Cerridwen, the northern goddess Wyrd, the Hindu goddess Kali, the Mycenaean Demeter, or the Babylonian Siris, the Goddess stirs her menstrual (or sometimes milky) womb/cauldron to generate, destroy and regenerate all living things.[321] In this great steaming menstrual soup all that will be curdles, coagulates and churns. The cosmogonic cauldron's undifferentiated, unclotted elements of possibility without form are probably alluded to in Genesis 1.2, where **tohu** and **bohu** are often translated perjoratively as the mere 'waste and void' of the primordial state. Her creation has no beginning and will have no absolute end. It is a continual cycle: she bears all the temporal aspects of Virgin, Mother, Crone at once. Change is an inseparable part of the meaning of the Goddess' existence, hence the recurrent crescent moon, chrysalis and butterfly symbols found in the archaeological remains of 'Old Europe'.[322] The chaotic cauldron metaphor is perhaps most significant … as it is fruitfully connected to those of new science and ecology, widely distributed in the world's mythologies, antithetical to the most harmful elements of the biblical cosmologies, and unsentimental in its image of both women and divinity.[323]

---

320 This is also discussed in Livingstone, *PaGaian Cosmology*, 57.
321 For details of the cauldron motif see Barbara Walker, *The Crone: Woman of Age, Wisdom and Power* (New York: HarperCollins, 1988), 92-122.
322 See Gimbutas, *The Goddesses and Gods of Old Europe*, 237.
323 *Thealogy and Embodiment*, 269.

It is an "ancient cosmology in which chaos and harmony belong together in a creation where perfection is both impossible and meaningless;"[324] yet it is recently affirmed in Western scientific understanding of chaos, as essential to order and spontaneous emergence.

Related to the clarifying and grounding of 'conception,' is the clarifying of the significance of the female egg, the ovum, to the process. Fertile things have been described as "seminal" for ages, along with the story that the male contributed the *seed*, planted *his seed* in the female: that his sperm was the seed … as Rachel Adler pointed out,[325] in the Western religious traditions conception was valued as the essential creative act and was seen largely as a male accomplishment. The female was figured as passive receiver, providing the inert dirt for the seminal sowing. Western science did eventually 'discover' the female egg in 1827, and it is known that the ovum is actually the seed: thus "seminal' is a female-referring power/metaphor. Ancient and indigenous cultures had no problem with understanding this, "a female-centred conceptual space,"[326] as can be witnessed in ancient iconography such as Demeter handing knowledge (as grain/seed) to Her daughter Persephone. But the old patriarchal story continues to be told and affective in many mainstream texts. I have suggested describing fertile things as "ovarian," to assist in making the shift; others make a case for reclaiming "seminal" as a female-referring quality, while another possibility is the word "germinal."

"the veil is thin that divides the worlds"

This is traditionally stated in regard to the Season of Samhain/Deep Autumn in particular, and also felt to be true of Beltaine, its opposite on the Wheel. The manifest and the manifesting, form and formlessness, are always in movement, re-cycling constantly: there is no "away." Light and dark are held in one embrace, as 'seen' in the sphere of the Moon – simply always moving. Rigid hard boundaries are an

---

324 Ibid., 270.

325 Rachel Adler, "A Mother in Israel," in *Beyond Androcentrism: Essays on Women and Religion*, edited by Rita Gross, 237-255 (Montana: Scholar's Press, 1977), 241.

326 Irene Coates, *The Seed Bearers – role of the female in biology and genetics* (Durham: Pentland Press, 1993: xv.

illusion at any time, but it is recognised more in this season, between the old and the new, as the dark continues to wax. These two modes of being, the dark and the light, the imaginal and the manifest, are always in a dance and movement. Vicki Noble in *The Double Goddess* refers to a book on pre-patriarchal India where dual sister deities inhabit the same yoni: "Light and dark are not opposing forces but transforming and revolving halves of the same wheel.[327] In the Celtic tradition, Great Goddess Brigid/Bride is often referred to as a duality: the new young one with the old aspect of *Cailleach,* "reflective of life and death, day and night, summer and winter, and perhaps of positive and negative, north and south."[328] The indigenous mind knows this connection: that the hard edges of things are an illusion of our visual field.

> Indigenous societies intentionally build living relationships with these energies so that death and life are not severed. In this way they maintain the balance that is at the heart of their ontology, one that perceives the environment as both unseen and physical as part of an unbroken cycle.[329]

Marija Gimbutas also notes this as the main theme of Goddess symbolism: the continuity of birth and death and the renewal of life "not only human but all life on earth and indeed in the whole cosmos."[330] Russian scientist Vladimir Vernadsky has described:

> At each moment there are a hundred million million tons of living matter in the biosphere, always in a state of movement. The mass is decomposed, forms itself anew mainly by multiplication.

---

[327] 6 referring to Gita Thadani, *Sakhiyani: Lesbian Desire in Ancient and Modern India* (London: Cassell, 1996), 20.
[328] Stuart McHardy, "Bride in Scotland," in Patricia Monaghan and Michael McDermott, eds., *Brigit: Sun of Womanhood* (Nevada: Goddess Ink, Ltd., 2013), 54.
[329] Elisabeth Sikie, *Patterns of Invocation in Neolithic Art* (2004-2005).
[330] See *The Language of the Goddess*, xix.

> Generations are thus born … unceasingly.[331]

and Vernadsky also describes:

> The biosphere consists of an enormous mass of matter, of which less than 1 per cent is in the 'activated' form of living matter, the recipient of the Sun's energy.[332]

And in this magic place, where the "veils are thin," we may take the stories in our hands, co-create new stories, be part of the change we want. A Celtic poem from the sixth century states:

> I have flown as an eagle,
> Been a coracle on the sea,
> I have been a drop in a shower.
> A sword in hand,
> A shield in battle,
> A string in a harp,
> Nine years in enchantment,
> In water, in foam,
> I have absorbed fire,
> I have been a tree in a covert,
> There is nothing of which I have not been a part.

And the journey may be a shamanic one

The journey into the dark sentience, into the 'manifesting,' into the "Dreamtime" as Starhawk calls it in her story,[333] may be a shamanic journey: that is, into the unseen realm where life is renewed. By shamanic, I mean that individuals are

> "knowing themselves and thinking from within their own skin. Shamanism relies on direct lived experience for an understanding of the sacred, as opposed to relying on an external authority,

---

331 Vladimir Vernadsky, *The Biosphere* (London: Synergetic Press, 1986), 34.

332 Ibid., 41.

333 *The Spiral Dance*, 257.

> external imposed symbol, story or image. Each person must claim their own inner power, imagine or visualize themselves, and use this in the service of life. … Religion then becomes based on what we can feel, what we can know. We each then find 'for ourselves our individual role in the matrix,' a way in which each being is part of the texture of the universal fabric. We are linked by a recognition in each other of a power that arises from within, not by some external word."[334]

The Creative Dynamic happens within each being; we are all the Mother, co-Creators with Her. Her changing faces are present in each being: all and each are She – She Whose face changes. We are all journeyers – *Journeyers* … the Seeds of our own rebirth.

The whole dark phase of year's wheel is a celebration of the Crone/Old One qualities but particularly Lammas and Samhain. Lammas may be understood as an *identification* with the Crone/Old One as we dissolve into Larger Self – Her Soup. We step into the power of the dark at Autumn Equinox – its grief and loss, and Samhain may be understood as a participation in Her *process* of the end of things and transformation. The Crone/Old One is that movement back into the Great Sentience out of which All arises. The Crone aspect is the contraction if you like, that may be understood as a systole, a contraction of the heart by which the blood is forced onward and the circulation enabled.[335] She is the systole that carries All away – She is about loss, but the contraction of the heart is obviously a creative one, it is the pulse of life. Perhaps it is only our short-sightedness that keeps us from seeing contraction/death that way: we insist on taking it all so personally, when it is not, in fact.

---

334 Glenys Livingstone, "Feminism and the Future of Religion," a paper presented at the National Socialist Conference in Sydney 1990, published on-line in 2014 https://pagaian.org/2014/04/06/feminism-and-the-future-of-religion/, quoting within it Neville Drury, The *Elements of Shamanism* (Element Books: Dorset, 1989), 101.

335 I owe the "systole" metaphor to Loren Eiseley, *The Immense Journey* (New York: Vintage Books, 1957), 20. He uses it to speak of an eternal pulse that lifts Himalayas, and then carries them away.

The falling apart, the disintegration can be, and usually is, felt as a scary or depressing thing. It is the loss of the known, perhaps the loss of a worldview, the loss of a world. Samhain is a time for such reflection – facing our deep fears and contemplating our demise. We usually confront fears or depression when we feel that our lives are changing – the world is falling apart. Samhain is an opportunity for immersion in a deeper reality which the usual cultural trance denies. It may celebrate immersion in what is usually 'background' – the real world beyond and within time and space: which is actually the major portion of the Cosmos we live in.[336] Samhain is about understanding that the Dark is a fertile place: in its decay and rot it seethes with infinite unseen complex golden threads connected to the wealth of Creativity of all that has gone before – like any compost. So, it is in the current dark times culturally, and so it may be – we may be some of the *Golden Threads*. We must retrain our perception to see the beauty of new form possible in the old decaying fermenting compost. While our ancestors may have wondered whether they would make it through the coming Winter, we may wonder in our times whether we will make it through the global spasms and planetary changes – and in what form.

Life will persist … such hope is represented by the Seed in the Fruit (as in an apple cut across the diameter). There is a thread of life that has eternally endured. There is the Seed within us, the Dark Shining One, already present, simply awaiting its time of awakening. It may not be in the form we had in mind. To perform the magic, we accept all the old shapes of our lives, and ultimately of our culture, as compost for the new. We accept the unknowing. We put it all in the pot, consume it all … feed our imaginations good food so that the new may emerge. We may make "re-solutions." We transform the old into the new in our own bodyminds – we are the transformers. While it is true that we give ourselves over to the Great Transformer, the Old One's process – we are She, and thus we play our part in the transforming, in the imagining, in the manifesting.

---

[336] Modern astrophysics has found that "dark energy" makes up 73% of Universe and "cold dark matter" 23%, the ordinary matter of which we are made is 4%.

The Universe Story and Samhain

Our Place of Being has made amazing transitions, since the beginning in the Primaeval Fireball (or however you imagine/name it), with the aid of catastrophic events, and with a minimum of materials; and nothing is wasted – all becomes compost for renewal and manifestation. Remembering Gaia's story, and meditating upon it, gives humans some clue about the present, and perhaps how to proceed; and so it may also be with the remembering of our own personal stories, and cultural stories, and present planetary catastrophe.

At Samhain/Deep Autumn, where the story is one of journeying yet further into darkness, the 'Transformation of Death,' and therein the conception of the new, it is a time for celebrating the *becomings*, the unimaginable More that Gaia – as I name this Place – has become, and will still yet become. It is a time for remembering the ancestors – creatures, plant and human – out of whom we all, and the present have arisen; that we are the ancestors of the future, and that we may be completely free to imagine/conceive Much More … we may seed our imaginations with wisdom from the deep, and set in place the possibility of actions which create wholeness. We may name ourselves as "the Transformation of the Ages." We – collectively and personally – *are* the Ancient One. Brian Swimme says: "We are a mode of the evolutionary dynamics"[337] – we are Cosmogenesis: as opposed to "individuals." Samhain is a good time for contemplating the question "Who are you?," all the layers of that self, perhaps summarized as biological/historical, cultural/mythic and cosmological/unitive.[338]

And a quote from The Universe Story, that is resonant with Samhain:

> ... after billions of years of striving … Tiamat found herself pressed to the wall, exhausted by the effort, helpless to do anything more

---

[337] *The Powers of the Universe,* (DVD series 2005).

[338] This summary is resonant with Jean Houston's 3 layers/realms of self: (i) historic, factual, biological – "*This is Me*" – small, local self; (ii) mythic, symbolic – "*We Are*" – Larger Self; and (iii) the unitive – exists both beyond & within the other 2 – "*I Am*," *The Search for the Beloved: Journeys in Mythology and Sacred Psychology* (Los Angeles: Jeremy P. Tarcher, Inc., 1987), 23-28.

> to balance the titanic powers in which she had found her way. When her core had been transformed into iron, she sighed a last time as collapse became inevitable. In a cosmological twinkling, her gravitational potential energy was transformed into a searing explosion, ... But when the brilliance was over, when Tiamat's journey was finished, the deeper meaning of her existence was just beginning to show through.[339]

Notes on imagery and metaphors in the offered Ceremonial Script

Costume is an option: its purpose is to emphasize possibilities, to entertain your imagination, your dreams and wild places. The Cosmos Herself, and as Earth, has come through so many wild, amazing changes in Her thirteen point seven billion year story … was it possible to imagine what was next? In the earliest of Earth's days could the first cells have been imagined? Could the advent of eyes, or the music we now experience have been imagined? So too, in our particular lives, could we have imagined our own advents? Can we imagine the More, the Not-Yet culturally? We can at least allow the Space for it.

For "Casting the Circle"/"Creating the Sacred Space," the focus is on how ancient each element within us is, how we are its recycled presence, thus how old we each really are. This is a remembering that there is nothing we have not been. Evolutionary biologist Elisabet Sahtouris tells the story of one silica atom through all the evolutionary moments to illustrate this.[340]

Overall, in the dark part of the cycle, the elements may be acknowledged primarily in their collective aspect, how we are each a drop of, and inseparable from that collective, how we are elementally immersed in something much Larger – not individual at all, how we elementally belong to that "Soup." It may contrast poetically with the light part of the cycle, where there may be a focus on the sensual experience of the elements, a focus on the individual sensate knowing of the elements, and in this way knowing participation in All.

The Invocation is an invitation to name yourself "as you will":

---

[339] Swimme and Berry, *The Universe Story,* 60-61.

[340] See "Journey of a Silica Atom" by Elisabet Sahtouris https://thegirlgod.com/pagaianresources.php

this may be a recognition of the power and fertility of the imaginal space for conceiving a world, and also a recognition that all is possible. We may identify with the mythic and grow ourselves into larger aspects of ourselves. It may be an opportunity for you to express something about yourself that gets little 'airplay'; for example, I often do a modified version of my actual name, that expresses another realm or dimension of my being. You may choose to name yourself as you *are* named.

I like to use dark gold as a colour for fabrics and ribbons for this Season, with the understanding that *Gold* is the colour of the transformation process. Samhain is about celebration of the process of transformation that takes place in the dark chrysalis. If we have the eyes to see from a deeper larger perspective, this is what death is: transformation. The Egyptian Goddess Selket, known to the ancients as a Goddess of transformation, is golden: the ancients seem to have known something that we have forgotten and are trying to remember.

Some of the format of this celebration is based on part of a poem by Robin Morgan,[341] with slight rewording. The metaphor of the snake is used because it is a representation of renewal with its shedding skin. Robin Morgan says:

> *Drawn from the first by what I would become,*
> *I did not know how simple this secret could be.*
> *The carapace is split,*
> *The shed skin lies upon the ground.*
> *I must devour the exoskeleton of my old shapes,*
> *wasting no part…*[342]

In this process of transformation Gaia wastes nothing.

One of the main rites is a remembering process in three parts: (i) we remember our old selves – who we have been (ii) we remember the dead – our ancestors and those who have gone before (iii) we remember the old cultural selves – collective – what humans (of any

---

341 Robin Morgan, "The Network of the Imaginary Mother." In *Lady of the Beasts* (New York: Random House, 1976), 61-88.
Copyright 1974 by Robin Morgan, reprinted with permission from *Lady of The Beasts* by Robin Morgan (Doubleday).

342 Ibid., 84.

culture) have done, what we are doing, some of the horrors … the ongoing wars, the inquisitions, the holocausts, the genocides, the geocide. We recall the ancestors, selves and cultural/historical events that have brought us to this moment.

The remembering of old selves is enabled with the playing of a children's game, "In and Out the Windows," where each participant travels in and out of upraised and linked hands and arms of the circle, and when "In" may speak and/or show photos of themselves from the past. Some may choose to remember any self from the entire evolutionary story, with whom they would like to identify. Each is praised for their "becomings," and all are praised as "Great Ones" for having come through so many changes/transformations as Gaia Herself has done. They are each presented with gingerbread snakes, "Gaian totems of life renewed," which will be consumed in three parts. See Appendix B for a recipe.

For the rebuilding/conceiving the new, the metaphor of the spider's thread is used because she weaves her home from her own bodymind with the strongest natural fiber known. The spider is a great example of Creator. As Robin Morgan continues (after the devouring of old shapes):

> *… free only then*
> *to radiate whatever I conceive,*
> *to exclaim the strongest natural fiber known*
> *into such art, such architecture*
> *as can house a world made sacred by my building.*[343]

The strongest natural fiber known is perhaps our imaginations, our creative selves. Samhain is the time for conceiving and imagining the Not-Yet – for dreaming the new. Garden string or a golden thread may be wrapped or "spun" in some way in the ceremony. In the offered ceremonial script, each participant wraps the string/thread around themselves and passes it to the next. The connected participants then make a journey, "Sailing to the New World," "across the Womb of

[343] Ibid.

Magic and Transformation … to the Not-Yet who beckons."[344]

Scissors are used in the ceremony to cut the thread or string, that represents each other's conceptions. Scissors are one of the tools of the Crone, for the finishing; which is at the same time the creating. This is one of her sacred processes, that all constantly participate in … and we may do so consciously.

Apples and apple juice are the chosen food for the Communion rite, as the apple in this Goddess tradition represents never-ending renewal.[345] In the tradition as Starhawk does it, an apple is frequently ritually cut at this Season's celebration to reveal the pentacle core and the seed in the fruit;[346] it suggests how we are held in the Heart of the Mother.[347] Consuming the apple as a holy food is an opportunity to re-story it, given that many participants will have been imbued with the Judeo-Christian story, wherein the apple is a fruit of temptation representing knowledge that is ruinous. Whereas the ceremonial eating of apples and drinking apple juice, may be understood as enjoying the fruit of never-ending renewal and deeper self-knowledge in the Heart of the Mother.

---

344 Brian Swimme speaks of how the "Not-Yet" reaches into the present, as an attractor; how the human imagination may be a space for shaping the future – personally and collectively. See *The Earth's Imagination*, program 8, "The Surprise of Cosmogenesis."

345 See Walker, *The Woman's Encyclopedia of Myths and Secrets*, 48-50.

346 Starhawk, *The Spiral Dance*, 212.

347 See Walker, *The Woman's Encyclopedia of Myths and Secrets*, 49.

[Figure 10] A Samhain/Deep Autumn Ceremonial Altar

A SAMHAIN/DEEP AUTUMN CEREMONIAL SCRIPT

Note that the directions here are called in "counter-clockwise" direction which is sunwise in the Southern Hemisphere. The order may be changed to "clockwise"/sunwise for the Northern Hemisphere. Note that this particular offered ceremonial script is not considered prescriptive, and that there are many creative possible variations.

Some requirements and some suggestions: Participants may come dressed in costume. Each bring photos/objects that represent "old selves" to have with them in circle. Covered basket of gingerbread snakes (see Appendix B for a recipe). Sliced apples (rubbed with lemon.) Apple juice poured in small glasses. Bowl of sunflower seeds, candy snakes. Ball of thread/garden string. Scissors. Photos of ancestors and passed loved ones for side altar/table. Black centre altar cloth with gold stars/webs/threads in it, or a dark golden cloth. "Mists"/veils of varied colours as decorations. Music ready.

Call to Gather – drumming until all are gathered or a little longer.

Centering – Breath Meditation: Welcome to all you magical creatures and beings, who are slipping in from your dreams and wild places – the possibilities between the worlds. Breathe, go within for a moment.

Breathe deep. … as you let go of your breath, follow it down, and notice the Space between your breaths. Don't hold it, just notice it. And again, breathing deep and following the breath down to this Space – that is always with you, feeling it. Breath deep – and as you let go, noticing this Space … into which all who have gone before us have travelled … from which to enter again the dance of life in some form. Feel this Space between your breaths.

Statement of Purpose: We are gathered to celebrate Samhain. This is the time when we recognize that the veil is thin that divides the worlds. It is the New Year in the time of the year's death – the passing of last Summer's growth. The leaves are turning and falling, the dark continues to grow, the light part of the day is getting shorter and colder. Earth's tilt continues to move us away from the Sun.

The story of old tells us that on this night, between the dead and the born, between the old and the new, all is possible; that we travel in the Womb of the Mother, the Dark Shining One within, from whom all pours forth, and that we are the seed of our own rebirth. The gates of life and death are opened: the dead are remembered, the Not-Yet is conceived. We meet in time out of time, everywhere and nowhere, here and there ... to transform the old into the new in our own bodyminds.

Let us enter this realm, the vast sunless sea within us – the Womb of All. We may proceed by remembering our elemental origins.

Calling the Directions/Casting the Circle

Celebrant: Hail the East, Powers of Water. We remember that we are Water.

All: We remember that we are Water.

Celebrant: Water, that has nursed our beginnings in the primordial soup, that walks around in our flesh. We are old, so old.

All: We are old, so old.

.... slowly walk the circle spooning water w/ shell, repeating: "we are old, so old."

AFTER WATER IS PUT BACK ON ALTAR, all: "We remember that we are Water & there is nothing we have not been."

Celebrant: Hail the North, Powers of Fire. We remember that we are Fire.
All: We remember that we are Fire.
Celebrant: Fire, that has surged through every thought, and action, that dances at the root of all life. We are old, so old.
All: We are old, so old.
.... light fire and slowly walk the circle with the firepot, repeating: "we are old, so old."

AFTER FIRE IS PUT BACK ON ALTAR, all: "We remember that we are Fire & there is nothing we have not been."

Celebrant: Hail the West, Powers of Earth. We remember that we are Earth.
All: We remember that we are Earth.
Celebrant: Earth, whose intelligence has conceived us and all creatures, and to whom we all return. We are old, so old.
All: We are old, so old.
.... slowly walk the circle with the rock, repeating: "we are old, so old."

AFTER EARTH IS PUT BACK ON ALTAR, all: "We remember that we are Earth & there is nothing we have not been."

Celebrant: Hail the South, Powers of Air. We remember that we are Air.
All: We remember that we are Air.
Celebrant: Air, that passed through the lungs of ancestors, dinosaurs, and every breathing creature, and inspires us now in every moment. We are old, so old.
All: We are old, so old.
.... light smudge, slowly walk the circle w/ smudge and feather, repeating: "we are old, so old."

AFTER SMUDGE IS PUT BACK ON ALTAR, all: "We remember that we are Air & there is nothing we have not been."

Celebrant: We are this Mystery – (PACING THE CIRCLE) Water, Fire, Earth and Air. We are from all time and no time, everywhere and nowhere. Feel the space within you … the Womb of All, fertile with possibility. Take a moment with Her now, breathe … This is the Centre. The Centre of the Cosmos is here (gesturing to the centre).

The circle is cast, we are between the worlds, beyond the bounds of time and space, where light and dark, birth and death, joy and sorrow, meet as One.
(Celebrant light centre candle.)

Invocation
Celebrant: We are Sacred Ones – divine, magical co-creators. We call the Divine by a thousand names, uttering ourselves. Name yourself as you will.
Each one in turn: I am ....
Group response: Deep bow & "Welcome ……. (full name as announced)."
Celebrant: We welcome all these Divine/Sacred Ones – magical Co-creators.

Transformation Journey – "In and Out the Windows"[348]
Celebrant: Let us now remember some the old selves we have been. Let us each take the journey now, remembering some of our transformations. THE CIRCLE TAKES HANDS
The circle raises hands, each one goes in and out (anti-sunwise – we are going back in time), while the circle chants:
"in and out the windows, in and out the windows, in and out the windows, as we have done before," then bringing their hands down, allowing the person to be in … and the chant stops.

---

[348] This process is an adaptation of a children's game as noted earlier. It seems appropriate to what each being does existentially in so many ways, over the eons as well as in our personal lives. The chant can be found in part at https://thegirlgod.com/pagaianresources.php.

All ask: "Who have you been?"
The person may show a photo/object – around the circle (walking around if they like), saying "this is someone I have been"; and/or they may choose to tell a story/stories about "old selves."

Group response: "Hail to you and your becomings."
Receive the greeting with a bow.
Each may do this up to three or four times as they wish.
Each return to circle when done – put any photos/objects on altar.

After all have woven in and out.

Presentation of Snakes
Celebrant (taking basket of snakes & walking the circle):
"O Great Ones: you are Great Ones, courageous ones. You have come through so many changes (indicating photos/objects on altar), ... as Gaia Herself has done. Gaia, like you, has come through so many changes/transformations – so many old selves. (elaborate with DRAMA)
O Great Ones, so many changes you have come through – as Gaia Herself has done.
… and yet – and YET: She and you – ***we*** – are More, much more."
UNCOVER SNAKES
"Accept these snakes, Gaian totems of life renewed – as you, like Gaia, have done so many times … and will yet again. You and Gaia, are More, much More."

Celebrant serve to one person repeating the blessing and naming: ."…. accept this Gaian totem of life renewed – you are More, much more."
Response: "It is so."
Each serve person sunwise around the circle repeating the blessing and naming: ."…. accept this Gaian totem of life renewed – you are More, much more."
ALL SIT

Consuming Old Selves
"Drawn from the first by what you might become,
You did not know how simple this secret could be.

The carapace is split, the shed skins lie upon the ground,
(hold up snake)
Devour now all your old shapes, wasting no part."

Celebrant breaks off a third of her snake
another echoes: "I devour now all my old shapes, wasting no part."
All echo and continue: "I devour now all my old shapes, wasting no part."
DRUM OR MUSIC OF CHOICE
All eat a third of their snakes.

Remembering the Ancestors
"Let us remember our ancestors, those who have gone before, whose lives have been harvested, whose lives have fed our own."
.... each may name those who have died, perhaps in a litany style.

After all have spoken, celebrant: "We welcome all these, whose lives have been harvested, whose lives have fed our own. We remember that we too will be consumed, feed others with our lives. May we be interesting food. We become the ancestors. We **are** the ancestors."
Celebrant hold up her snake and break off a part, and says:
"We become the ancestors. We are the ancestors. We will be consumed."
another echoes: We become the ancestors. We are the ancestors. We will be consumed.
All echo and continue: "We become the ancestors. We are the ancestors. We will be consumed."
DRUM OR MUSIC OF CHOICE
All eat the more of their snakes.

Remembering the Old Shapes of the Culture
"Let us remember some of the old shapes of our human culture and story that we would leave behind, that we might transform in our own bodyminds."

.... each name the "old shapes"/stories they choose
After all have spoken, celebrant hold up remaining bit of her snake: "We devour these old shapes, transform them in our beings."

Another echoes: We devour these old shapes, transform them in our beings. All echo and continue: "We devour these old shapes, transform them in our beings."
DRUM OR MUSIC OF CHOICE
All eat the rest of their snakes.

Sitting in Silence
Celebrant: Let us sit for a while with all these endings.

Building the Web – Conceiving the Future
Celebrant takes ball of thread

"Having devoured these old shapes, wasting no part
we are free … free to radiate whatever we conceive,
to exclaim the strongest natural fibre known
– our creative selves,
into such art, such architecture
as can house a world made sacred by our building."
"Take the thread now, wrap it around, and let us build a world of our conceiving."
ALL STAND
Celebrant holds the end of the ball of thread, wraps it around herself and passes the ball sunwise, with all chanting:
"Free to radiate whatever we conceive."
until all hold the thread (last person has ball).

Sailing to the New World
MUSIC ON LOW (some "journey" music)[349]

Celebrant: Let us set sail now for a new world – beyond our knowings, across the vast sunless sea between endings and beginnings … across the Womb of Magic and Transformation … to the Not-Yet who beckons. As the New Year turns – the New Era turns."
Celebrant (switch music up), and lead group in a circle around the centre

---

[349] I have used the first part of *1492* by Vangelis (1992) or *El Condor Pasa*, Daniel Alomia Robles (1968).

altar sunwise first, then folding back on the circle, then around into the dark (another nearby space?) repeating text once or twice in the process, perhaps humming, swaying. Then lead **into a spiral.** At centre of spiral: "to the hinge of the spiral where past present and future meet, where what has been consumed can be renewed … where All Possibility is quickened to new life by what has been."[350]
AS MUSIC ENDS: "stepping ashore the Shining Isle – the Luminous Egg."

When back in circle, after the journey:
Celebrant take the ball of thread from the last person, and wraps it around herself, completing the circle of thread: "On this Shining Isle, in this magical Place, we are free to radiate whatever we conceive. What would you conceive, imagine, create?

Celebrant passes the ball sunwise for each in turn to hold – and wrap around themselves as they say:
Each: "I conceive/spin/imagine/create … "
Group response: So be it!
Each cuts both threads for next person: "You are free – magical co-Creator."
Pass scissors – each cut for next.

Celebrant: "May all these conceptions, desires, imaginings – spoken & unspoken – create a world made sacred by our building."
All: May it be so.
Celebrant put thread down.

Communion
Celebrant hold up plate of sliced apple & glass of juice.
"Stand tall, Daughters and Sons of the Mother. Daughters and Sons of Eve, stand tall – eat and drink the fruit of never-ending renewal and self-knowledge."
Another person hold tray of glasses of poured juice

---

[350] A paraphrase of Starhawk's words, *The Spiral Dance*, 236-237.

Celebrant offer juice and apple slices, repeating the blessing – "Stand tall daughter/son of the Mother – eat and drink the fruit of never-ending renewal and self-knowledge."
GROUP TOAST
Offer more apple, cider and juice, lolly snakes and seeds.

Story Space (with ceremonial manners)[351] – stories of transformation.

Open the Circle
Celebrant: We have remembered this evening that we are Air, as old as She, and present in the lungs of all who have ever breathed.
May there be peace within us.
All: May there be peace within us.

Celebrant: We have remembered this evening that we are Earth, as old as She and conceived of Her intelligence.
May there be peace within us.
All: May there be peace within us.

Celebrant: We have remembered this evening that we are Fire, as old as She, and surging with Her dance.
May there be peace within us.
All: May there be peace within us.

Celebrant: We have remembered this evening that we are Water, as old as She, and nursing all possibility.
May there be peace within us.
All: May there be peace within us.

Celebrant: We have remembered this evening that we are Sacred Ones – divine, magical co-creators. We have remembered some of our transformations, and those who have gone before us. We have remembered old shapes of our human culture and story that we would

---

[351] These manners require listening with no comment or additions to the speaker's story, and confidentiality.

transform in our bodyminds, and we have conceived a world made sacred by our building. We have remembered that we are More, much More – as Gaia Herself is. May there be peace within us and between us. (take hands).
All: May there be peace within us and between us.

Celebrant: May the peace and bliss of the Creative One go in our hearts. The circle is open, but unbroken.

All: It has been a merry meeting
It is a merry parting,
May we merry meet again.
Blessed Be!

Points for individual contemplation prior to the ritual:

- Naming yourself at the Invocation.
- Old selves you have been.
- Ancestors or those who have gone before whom you would like to name.
- Old cultural shapes, stories, events you see yourself as helping transform.
- What you would like to create in your life … conceive, imagine?

Samhain/Deep Autumn Goddess Slideshow:
https://thegirlgod.com/pagaianresources.php

CHAPTER 5
WINTER SOLSTICE/YULE

Southern Hemisphere – June 20 – 23.
Northern Hemisphere – December 20 – 23

In this cosmology Winter Solstice is particularly a celebration of *She Who is this Dynamic Place of Being*, the Mother/Creator aspect of the Triple Goddess – as both Solstices are. This aspect of the Creative Triplicity is associated with the *communion* quality of Cosmogenesis.[352] This Moment is the ripe fullness of the Dark Womb and it is a gateway between dark and light. It is a *Birthing Place* – into differentiated being. Whereas Samhain/Deep Autumn is a dark conceiving *Space*, it flows into the Winter Solstice dark birthing *Place* – a dynamic *Place of Being*, a *Sacred Interchange*. This Seasonal Moment of Winter Solstice is the peaking of the dark space – the fullness of the dark, within which being and action arise. It is the peaking of emptiness, which is a fullness. As cosmologist Brian Swimme describes: the empty "ground of being … retains no thing." It is "Ultimate Generosity."[353]

In Vajrayana Buddhism, Space is associated with Prajna/wisdom – out of which Upaya/compassionate action arises. Space is highly positive – something to be developed so appropriate action may develop spontaneously and blissfully.[354] In Old European Indigenous understandings, the dark and the night were valued at least as much as light, if not more so: time was counted by the number of nights, as in 'fortnights,' and a 'day' included both dark and light parts … it was 'di-urnal'. I have been careful with my language about that inclusion in the ceremonial 'Statement of Purpose' for each Seasonal Moment. This awareness is resonant with modern Western scientific perceptions about the nature of the Universe: that it is seventy-three percent "dark energy," twenty-three precent "dark matter," four percent

---

352 Swimme and Berry, *The Universe Story*, 77 – 78.
353 Swimme, *The Universe is a Green Dragon*, 146.
354 See Rita Gross, "The Feminine Principle in Tibetan Vajrayana Buddhism." *The Journal of Transpersonal Psychology,* Vol.16 No.2 (1984), 179-192.

"ordinary matter."[355] The truth is that we live within this darkness: it is the *Ground of all Being*.

Some Winter Solstice Story

This is one of the easiest of Earth's holy days for people of our time in general to relate to, particularly in Christianized cultures, where it has been celebrated as "Christmas" since the Middle Ages. It is after all the birth of Light that is celebrated! The Winter Solstice marks the stillpoint in Winter, when Earth's tilt causes the Sun to begin its return. It may have been the first Earth-Sun event that the ancients noticed, it seems the most obvious and dramatic; when the region has been drifting into coldness and darkness, and then that stops and turns – what a relief. And it is this Seasonal Moment which the pre-Celtic ancients apparently particularly celebrated with the construction of Brú-na-Bóinne, known as "Newgrange" since its re-discovery in the Christian era. The major structures there were completed between 3200 and 3700 B.C.E..[356] Brú-na-Bóinne – Newgrange – had been thought to be a tomb, and some still refer to it as such. But more recent research, primarily initiated by Martin Brennan late last century, and documented in his book *Stones of Time*, reveals something else – a large scale solar-construct: a "clock" for telling time.

It is at the Winter Solstice that the inner chamber wall of Brú-na-Bóinne carved with the Triple Spiral, is lit up by the rising Sun. So, in PaGaian Cosmology special note is made of this motif at Winter Solstice, and an attempt is made to plumb the depths of understanding it, and re-imagining its significance. One of the primary ceremonial processes as it has been done for PaGaian Winter Solstice is the "Cosmogenesis Dance," which may be understood as a celebration of the Triple Spiral dynamic. This dance has been known as the "Stillpoint Dance";[357] yet after doing it for a few years I noticed its three layers as resonant with the Triple Goddess. I then re-named it, and developed consciousness of it as a specific celebration of the Triple Spiral, and thus Cosmogenesis.

---

[355] These figures as told by cosmologist Paul Davies with Macquarie University's Centre for Astrobiology, Australia.

[356] See Brennan, *The Stones of Time.*

[357] And I learned it originally by that name from Jean Houston in 1993.

The Triple Spiral is a celebration of Earth-Sun relationship in some way … some say it immortalizes "earth's power to give new life to the sun,"[358] and some say it is the other way around.[359] But there has been little awareness until recent times of the Indigenous "Goddess" mind required for comprehending it. One might say now that it celebrates both powers: that is, the *relationship*, and *apparently* it celebrates the *triplicity* of the Cosmic Dynamic particularly at this Seasonal Moment. Winter Solstice may be expressed as a gateway from Crone/Old One (dark) through Mother to Virgin/Young One (light) – the birth of form, of life … this is how it happens. Nothing is excluded from these three phased dynamics – not Earth, nor Sun, nor you or me, any being – female and male and all genders/sexes. It is an "innate triplicity of the Cosmos … that runs through every part of the Universe."[360] Winter Solstice is a good moment to recall and celebrate this seamless connection.

In Pagan traditions since Celtic times, and in many other cultural traditions, Winter Solstice has been celebrated as the birth of the God; and in Christian tradition since about the fourth century C.E., as the birth of the saviour. But there are deeper ways of understanding what is being born: that is, who or what the "saviour" is. In the Gospel of Thomas, which was not selected for biblical canon, it says: "If you bring forth what is within you, what you bring forth will save you."[361] This then may be the Divine Child, the "Saviour": it may be expressed as the new Being forming in the Cosmogonic Womb,[362] who will be born. We may celebrate the birth of the new Being, which /who is always beyond us, beyond our knowing … yet is within us, burgeoning within us – and within Gaia. What will save us is already present within – forming within us. The Winter Solstice story may emphasize that what is born, is within

---

[358] Elinor Gadon, *The Once and Future Goddess* (Northamptonshire: Aquarian, 1990), 346.

[359] Baring and Cashford, *The Myth of the Goddess*, 98.

[360] Matthews, *The Celtic Spirit*, 366.

[361] Elaine Pagels, *Beyond Belief: the Secret Gospel of Thomas* (NY: Random House, 2003), saying number 70. See https://www.pbs.org/wgbh/pages/frontline/shows/religion/story/thomas.html.

[362] Melissa Raphael's term, *Thealogy and Embodiment,* 262.

each one – the "Divine" is not "out there": it may be said, and expressed ceremoniously, that we are each *Creator and Created.*

We may imagine ourselves as the in-utero foetus – an image we might have access to these days from a sonar-scan during pregnancy. This image presents a truth about Being: we are this, and it is within us, within this moment. Every moment is pregnant with the new. It will be birthed when holy darkness is full. Part of what is required is having the eyes to see the "new bone forming in flesh," scraping our eyes "clear of learned cataracts,"[363] seeing with fresh eyes. That is what the fullness of the Dark offers – a freshening of our eyes to see the new. And the process of Creation is always reciprocal: we are Creator and Created simultaneously, in a "ngapartji-ngapartji"[364] way. We are **in-formed** by that which we **form**. In Earth-based religious practice, the ubiquitous icon of Mother and Child – Creator and Created – expresses something essential about the Universe itself … the "motherhood" we are all born within. It expresses the essential *communion* experience that this Cosmos is, the innate and holy *Care* that it takes, and the reciprocal nature of it. We cannot **touch** without being **touched** at the same time.[365] We may realize that Cosmogenesis – the entire Unfolding of the Cosmos – is essentially relational: our experience tells us this is so. The image of *The Birth of the Goddess* on the front cover of my book *PaGaian Cosmology* expresses that reciprocity for me, how we may birth each other and the healing/wholing in that exchange. It is a *Sacred Interchange.* And it is what this Event of existence seems to be about – deep communion, which both Solstices express.

Birthing is not often an easy process – for the birthgiver nor for the birthed one: it is a shamanic act requiring strength of bodymind, attention, courage, and focus of the mother, and resilience and courage to be of the new young one. Birthgiving is the original place of 'heroics,' which many cultures of the world have never forgotten, perhaps

---

363 The quotes come from a poem by Cynthia Cook, "Refractions," *Womanspirit* (Oregon USA, issue 23, March 1980), 59.

364 This is an Indigenous Australian term for reciprocity – giving and receiving at the same time. I explain it a bit further in *PaGaian Cosmology*, 256-257.

365 An expression from Abram, *The Spell of the Sensuous*, 68.

therefore better termed as "heraics." Patriarchal adaptations of the story of this Seasonal Moment usually miss the Creative Act of birthgiving completely, usually being pre-occupied with the "virgin" nature of the Mother which is interpreted as having an "intact hymen." The focus of the patriarchal adaptation of the Winter Solstice story is the Child as "saviour": even the Mother gazes at the Child in most Christian icons, while in more ancient images Her eyes are direct and expressive of Her integrity as Creator.

It should also be noted that in current Pagan traditions the Sun is frequently figured as male, but there is no necessity for this, though one may: for most of humanity's history and in many cultures, Sun was understood as Mother, not as a male principle.[366] Sun may be felt as Mother. And we are all children of Her. And over the years I changed language like "light" to "new being" because what I wanted to emphasize was the bringing forth of new being – out of the richness of the Dark, and "light" is commonly over-valued as a quality of being. This is how minds may be changed – with new Poetry, new words. Other language I have changed since publishing *PaGaian Cosmology* is the naming of the ceremonial process of "Re-Lighting the World" to "Re-generating the World." The Sun doesn't just "re-light the world" and that may not be felt as desirable … Sun "re-generates the world" – that is desirable.

Winter Solstice is the time for rejoicing in the awesome miracle of manifestation – at the beginning of time, and in every moment. It is a celebration of the Primeval Fireball – the Original Big Birth, as well as the birth of our Sun from the "Grandmother" supernova – whom Thomas Berry named as Tiamat,[367] and the birth of the first cell, and our own personal manifestation: a multilayered advent. It is the time for the lighting of candles, expressing what we wish to birth in ourselves in the coming year. It can be a moment for recalling the "Great Turning" of these times, as Joanna Macy names our era[368] – the hopes we might hold

---

366 See Monaghan, *O Mother Sun!*

367 Swimme and Berry, *The Universe Story*, 49 and 60-61. Tiamat was a Mesopotamian primordial sea Goddess, regarded originally as Creator. She is often imaged as a sea serpent/dragon. Thanks to Miriam Robbins Dexter for guidance on this understanding.

368 Macy and Brown, *Coming Back to Life*, 6.

for the future, and the actions we may take.

Winter Solstice is essentially a celebration of the Mother aspect of the qualities of Creativity; it is the ripening of Her darkness into the awesome act of creation of form, the web of life, the field of being. It is a celebration of Communion, a point of interchange from the "manifesting" into the "manifest"; it is a time for feasting and experiencing this essence of existence. At this point in the Wheel, She is the Alpha, and at the Summer Solstice She will be the Omega – both Gateways, points of interchange, when dark and light turn. At this Winter Gateway, the Crone's face passes through the Mother to the Virgin. Samhain, Winter Solstice and Imbolc, may be felt as the three faces of Cosmogenesis in the movement towards form.

Brian Swimme has said that it is the very essence of the Universe to give birth to the new.[369] The Universe is doing it all the time, and that is what 'allurement'/Allurement is all about – perhaps that is the "end of Desire' as it says in the Charge of the Goddess;[370] that is, to give birth to the new. And perhaps that is what the Triple Spiral represents. Swimme offers further, that to take on the mind of the Universe, is to become the "pulse of Creativity." This is Creator and Created; the ubiquitous icon of "Great Mother and Divine Child" is about the Creative impulse of the Universe, the end of Desire. We (all form and manifestation) are That. We affirm this and invoke this in each other in the Winter Solstice ceremony – we are this icon. In the ceremony offered, it is said to each: "Thou art Goddess Mother, thou art the Divine Child. Thou art all of That … a whole Universe"; and later, when each lights their candle: "May you bring forth what is within you."

Winter Solstice/Yule Breath Meditation

To begin this Seasonal Moment's contemplation:

> Centering for a moment … feeling the full cycle of your breath, and recognising it as the Triple Goddess Dynamic here in this primary place: receiving the Gift with the inbreath, feeling the

---

[369] *The Earth's Imagination,* video 7 "New Forms of Synergy."

[370] Doreen Valiente, referred to also in Chapter 2, quoted in Starhawk, *The Spiral Dance*, 102-103, and again in the context of Beltaine ceremonial script in Chapter 8.

waxing, feeling the peaking – the exchange, the communion that this Place of Being is, then the release – the dissolving, becoming the Gift, making space, ever transforming. This is the Triple Spiral dynamic here within you, in this primary place … feeling this breath.

Some Winter Solstice motifs

[Figure 11]

Suggested decorations for this Seasonal Moment, placed on the front door, windowsills, corners of shelves, and/or altars, may include:

- **maroon fabric** – representing dark ripe womb.
- **white ribbon** – representing birth of light and form.
- **pot for flame** (pot of sand which can be wet with methylated spirits and lit) – representing origins.
- **twig of evergreen and pinecone** – for eternal re-creation.
- **taper or tea candle** – representing the small self that is born.
- **image of in-utero babe** – we are always all this.

An Invocation

MOTHER[371]

Mother ... mother
Matrix of whom I am a part
Particles flux in and out
a continuation of the Original Ovulation.

In silence I reach to
increase the space inside of me.
As traction decompresses the spine
Silence and stillness decompress my spirit,
quickening new electro-chemical impulses
in my matrix ... matter ... mater
making room for new gestations
new birthings.
Mother,
Mother

reaching for you within me
reaching for me within you

Birthing you birthing me.

The Universe Story and Winter Solstice

One may contemplate the personal and collective primordial relationship with the Sun:

> How five billion years ago the hydrogen atoms, created at the birth of the universe, came together to form our great Sun that now pours out this same primordial energy and has done so since the beginning of time. How some of this sunlight is gathered up by the Earth to swim in the oceans and to sing in the forest. And how some of this has been drawn into the human venture, so that human beings themselves are able to stand there, ... are able to

[371] A poem by Glenys Livingstone 1995.

> think only because coursing through their blood lines are molecules energized by the Sun.[372]

One may contemplate the generous nature of this energy that fills all humans and beings:

> Human generosity is possible only because at the center of the solar system a magnificent stellar generosity pours forth free energy every day and night without stop and without complaint and without the slightest hesitation. This is the way of the universe. This is the way of life. And this is the way in which each of us joins this cosmological lineage when we accept the Sun's gift of energy and transform it into creative action that will enable the community to flourish.[373]

One may contemplate the Birthplace of the Universe:

> The idea that the universe began in one place is certainly an ancient one in human history. The image of a birthplace of the universe occurs in the mythical and classical forms of consciousness, and possibly even earlier. Such images as 'the cosmic egg' that cracks open and gives birth to all phenomena are found in Neolithic cultures around the planet.[374]

Thus one may contemplate the Western scientific theory that the universe began in an event of cataclysmic energy, an initial singularity of space-time, based on the observation that the clusters of galaxies are rushing away from each other, the universe is expanding, and has a "dual nature of being both old and new simultaneously."[375] In the 1960's two Western scientists, Arno Penzias and Robert Wilson detected the background radiation from the eruption of the universe: that is, they "discovered"/perceived the photons – the light particles – that

---

372 Swimme, *The Hidden Heart of the Cosmos*, 43.
373 Ibid., 44.
374 Ibid., 76.
375 Ibid.

originated in the primeval fireball itself, "the cosmic center, the world's navel, the sacred origin point of being. That is the place endowed with the stupendous fecundity necessary to give birth to the cosmos."[376]

Yet together with this is the realization that is implied in the scientific observation: that the superclusters of galaxies are all expanding away from where we the "observers" are, the *place* of observation, and thus also the apparent center of the birth of space and time. In brief, the implication is that we live in an omnicentric universe: that "in this universe of ours to be in existence is to be at the cosmic center of the complexifying whole."[377]

The Story of the Unfolding Universe is a story of EarthGaia's many Births

At Winter Solstice/Yule, it is the Original Flaring Forth, the Primaeval Fireball, the Great Origin, the Birth of All, that is echoed in the Sun's "return" (as it is said) and the movement out of darkness. Also echoed is the birth of our Solar System, from the Grandmother Supernova (who may be named *Tiamat* – a primordial Creator Goddess);[378] this is our particular Cosmic lineage whom we may remember at this time. This Cosmic lineage may be acknowledged as present in the births in our psyches, imaginations, and minds – as all are directly participant in the Cosmic Unfolding.

Since the Universe's story and thus that of Earth, is also your/our story, it seems important to find ways to learn it. My main sources have been *The Universe Story* by Brian Swimme and Thomas Berry, and *Earthdance* by evolutionary biologist Elisabet Sahtouris, but there is a growing body of resources for both children and adults, as the Western scientific version breaks into human awareness more deeply.[379] I also

---

[376] Ibid., 78.

[377] Ibid., 85-86. For more on the omnicentric nature of the universe see the whole chapter 80-89, and also this video excerpt "The BirthPlace of the Universe": https://thegirlgod.com/pagaianresources.php.

[378] "Tiamat" is Thomas Berry's name for the Supernova that was the Mother of our Solar System, *The Universe Story*, 49.

[379] Carl Sagan's work was groundbreaking in this regard, and there is Neil deGrasse Tyson's current work, and also the work of physicist

offer a "Cosmic Walk Script" which identifies just forty points of significance to contemplate, and the possibility of constructing a spiral walk for that purpose (see Appendix C).

And a quote from *The Universe Story*, that is resonant with Winter Solstice:

> A billion years after the birth of the universe, when the galaxies have just emerged, great regions of hydrogen and helium drift about the centre of the Milky Way. In the collapse of our galactic cloud, the spinning of the matter flattens out, disclike, as the angular rotation carries the clouds into the gentle movement of the twirling spiral galaxy. After another hundred million years the invisible density arm sweeps through the cloud and shocks it into collapsing upon itself. No further energy from the galaxy is now required. The cloud that has drifted undisturbed for eons suddenly undergoes a profound transformation that destroys its basic form but gives birth to a cluster of ten thousand diamond lights in a sea of dark night.[380]

Notes on imagery and metaphors in the offered ceremonial script

The initial chant and moving around the circle, together with expressive arm movements may encourage a recall of our galactic location in the Milky Way. It is also a good way to warm up, as it might be chilly. The joining of voices invokes the quality of communion being celebrated in particular.

The Breath Meditation begins with a focus on the 'void' at the bottom of the emptying of breath and feeling the urge to breathe as it arises. This focus is drawn into a recalling of the many birthings of all kinds in personal lives – remembering how one has been Creator and Created. That is then extended to recalling Gaia-Earth's many birthings in this moment, every day and throughout the eons, and then Gaia-Universe's many birthings in this moment, every day and throughout the eons. Attention then is again drawn back to the breath, associating it with

---

Brian Cox, and Jennifer Morgan who wrote an award-winning trilogy for children (or grown-ups), just to mention a few.

380 Swimme and Berry, *The Universe Story*, 47.

the birthing of all in every moment, the seamless Gaian self, specifically the seamless connection of the layers of being – of self, other and all: of self, Earth and Cosmos.

For "Calling the Directions/Elements"/Casting the Circle" the elements are remembered as Cosmic Dynamics, dynamics that translate into particular capacities within the human – sensitivity, shaping power, memory and wisdom, exuberance and expression: this is drawn from Brian Swimme's understandings of the elements.[381] Remembering them in this way at Winter Solstice extends the theme of this Moment's breath meditation as connecting the layers of the seamless Gaian self – our presence to Origins.

For the Invocation participants anoint each other with oil and pronounce with authority the Divinity/Sacredness in each other, expressing that each is Creator and Created: "Thou art Goddess, thou art the Divine Child, thou art all of That." These words are offered, and participants are encouraged to take this opportunity to "try them on," but also may be invited to choose to add something that they find expressive of the reality, like perhaps, "You are a whole Universe." This Invocation is extended then into the "Cosmogenesis Dance" which has three concentric layers to it and expresses in movement the three aspects of the Dynamic Cosmological Unfolding that we participate in. The dance begins with an inner and an outer circle, that move into relationship in a middle space, before passing into the opposite places – becoming outer and inner respectively. I identified its three interactive layers as embodying the qualities of (i) ever new differentiated being, (ii) the deeply related interwoven web, and (iii) the eternal creative return to All-That-Is, the hidden sentience within, the centre of creativity that each being is. The dance represents the flow and balance of these three – a balance and flow of Self, Other and All-That-Is. It may be experienced like a breath, that all breathe together: and in that process, all breath-taking manifestation co-creates the Cosmos. It is a process of complete reciprocity, a flow of Creator and Created, like a breath: there is dynamic exchange in every moment. The dance then may help participants to get it, to invoke it and to align themselves with it. (See Appendix D for Cosmogenesis Dance instructions and music).

---

381 Swimme, *The Universe is a Green Dragon*, 87-109 and 127-151.

Since we are celebrating the full darkness of the Mother, each participant is offered dark fruit cake and red wine/dark juice, and poetically identified as *Her* Cake and Wine – made by and for Her, integral with the Cosmic processes, affirming our Origin and our purpose, as we enter into the growing light part of the year. The naming of each participant as "Cake for the Queen of Heaven and Earth" is a reference from a biblical text (Jeremiah) where it is noted that the women are (still) baking cakes for the Queen of Heaven, Astarte – it's hard to stop a good thing![382] We are invoking an ancient practice, except that now in the current context, we are also acknowledging ourselves as the Cosmic cake. We *are* Her Communion, we are the Place/Space where it happens. This identification with the Communion food of cake and wine/juice also resonates with the Communion expression of Summer Solstice where participants affirm that they are "Food – Bread and Wine/Juice," that has ripened, ready for consumption.

---

[382] I first heard the term "Cakes for the Queen of Heaven" from the work of Rev. Dr. Shirley Ranck and adapted it. See https://www.cakesforthequeenofheaven.org.

[Figure 12] A Winter Solstice/Yule Ceremonial Altar

A WINTER SOLSTICE/YULE CEREMONIAL SCRIPT

Note that the directions here are called in "counter-clockwise" direction which is sunwise in the Southern Hemisphere. The order may be changed to "clockwise"/sunwise for the Northern Hemisphere. Note that this particular offered ceremonial script is not considered prescriptive, and that there are many creative possible variations.

Some requirements and some suggestions: ceremonial space lit with candles and "fairy" lights, and fire (for warmth) in a chimera. Evergreen wreath around a new centre candle and last year's old candle. Deep maroon altar cloth with white ribbon. Basket of dinner/taper candles and holders. Trays of sand on altar for placing lit candles. Decorated evergreen tree or trees. Bowls of oil. Cake and wine and juice. Since this particular offered script involves an inside fire: a fire cauldron of sand with methylated spirits for lighting (for a soft and safe flame), ready in centre of inside ritual space adjoining the ceremonial space. Solstice music playing during arrival time. Dance music ready. Copies of "Silent Night" and "PaGaian Joy to the World" to be handed out. (See Appendices E and F).

Call to Gather – drumming until all are gathered in circle.

Chant – with all moving around the circle in sunwise direction:

"A circle around, a circle around, the boundaries of the Earth.
A circle around, a circle around, the boundless Universe.
Spreading my long tail feathers as I fly,
spreading my long tail feathers as I fly,
Higher, higher, higher and higher.
Deeper, deeper, deeper and deeper."[383]
All join in movement and chant for some time, finishing eventually with each back in their place.

SITTING
Centering – Breath Meditation: Let us focus on our breath for a while. Take a deep breath and let it go. Notice the Void at the bottom of emptying your breath … feeling it, and feeling the Urge to breathe as it arises. And again … feeling it over and over – this breath that arises out of the full emptiness in every moment, birthing you in every moment.

– Recall some of the birthings in your life, your actual birth – see it there in your mind's eye … you coming into being – your *Nativity, your* Nativity. Recall projects you have brought into being, new beings within yourself, perhaps children, new beings in others, how you have been Creator and Created – even at the same time … who was birthing who? Staying for a while with the many, many birthings in your life.

– Recalling now Earth-Gaia's many birthings out of the Dark everyday … the dawn is constant as She turns. See Her in your mind's eye – the constant dawning around the globe, the constant birthing. Recall Earth's many births right now of all beings – as day breaks around the globe –

---

[383] I don't recall the source of this chant.

the physical, emotional, spiritual births. Her many, many birthings everyday, and throughout the eons.

– Recalling now Universe-Gaia's many birthings – happening in every moment – right now in real time and space … supernovas right now, stars and planets being born right now. Her many, many birthings in every moment and throughout the eons.

Come back to your breath – this wonder – none of it separate … the Origin Ever-Present, birthing you in every moment – out of Her Fertile Dark, in real time and space.

Feeling this breath, Her breath.

Statement of Purpose

This is the time of the Winter Solstice in our Hemisphere, and we are gathered to celebrate. Earth's tilt leans us away from the Sun to the furthest point at this time in our annual orbit. This is for us, the time when the dark part of the day is longest – darkness reaches Her fullness, She spreads her cloak, and yet gives way, and moves back into light. The breath of Gaia pauses. She rests. We wait ... within the Cauldron, the Dark Space, for the transformation.

The stories of Old tell us of the Great Mother giving birth to the Divine Child on this night. This Divine Child is the new being in you, in me ... is the bringer of hope, the evergreen tree, the return of warmth and light, the centre which is also the circumference – All of Manifestation. The Divine Child being born is the Miracle of Being, and the Unimaginable More that we are becoming.

Let us join the breath of Gaia in her suspension, the Great Mother in Her birthgiving. Let us recall this Dark Space, this Holy Cauldron within. Let us begin by remembering who and what we are and from whence we come.

ALL STANDING

Calling the Directions/Casting the Circle

All turning to the East

Celebrant: We come from the East to this Place, and we remember that

we are Water. Water that we are, we remember you.
All: Water that we are, we remember you.
Celebrant: Cosmic Dynamic of Sensitivity,[384] that absorbs, becomes, whatever you touch; may we *feel* what we are, and respond compassionately.
All: May we feel what we are and respond compassionately.
DRUMS as co-celebrant sprinkles water w/ pine branch.

All turning to the North
Celebrant: We come from the North to this Place, and we remember that we are Fire. Fire that we are, we remember you.
All: Fire that we are, we remember you.
Celebrant: Unseen Shaping Power of the Cosmos,[385] that gives us form – flames that we are; may we dance with you and act with Creative Lust for all of life.
All: May we dance with you and act with Creative Lust for all of life
DRUMS as co-celebrant lights fire in the pot, and takes it around.

All turning to the West
Celebrant: We come from the West to this Place, and we remember that we are Earth. Earth that we are, we remember you.
All: Earth that we are, we remember you.
Celebrant: Deep Sentient Presence and Memory,[386] you hold all the stories of life in your Body, we can learn it all from you; may we remember who we really are, may we hold the Wisdom of all time and no Time.
All: May we hold the Wisdom of all time and no Time.
DRUMS as co-celebrant holds up rock and carries it around.

All turning to the South
Celebrant: We come from the South to this Place, and we remember that we are Air. Air that we are, we remember you.

---

384 This understanding of Water comes from Swimme, *The Universe is a Green Dragon*, 87-95.
385 This understanding of Fire comes from Swimme, Ibid., 127-139.
386 This understanding of Earth comes from Swimme, Ibid., 99-109.

All: Air that we are, we remember you.
Celebrant: Cosmic Dynamic of Exuberance and Expression,[387] Wind that moves the trees, the clouds, brings us rain and allows us voice; move us and inspire us to unfurl our being.
All: Move us and inspire us to unfurl our being.
DRUMS as co-celebrant lights smudge stick and carries it around.

All turning to the centre
Celebrant: This is what we are – Water, Fire, Earth, Air (PACING THE CIRCLE) – Sacred Mystery, Cosmic Dynamics, manifest in this Place and Time. The Centre of Creativity is Here. The circle is cast, we are between the worlds, beyond the bounds of time and space, where light and dark, birth & death, joy & sorrow meet, as One.
(NOTE: the new centre candle will be lit later ceremoniously)

Invocation.
Celebrant: Let us now recognize and invoke this Sacred Creativity in each other.
Celebrant (taking bowl of oil and holding it up): As the oil comes to you, turn to the person next to you, anoint their forehead with oil, and pronounce with the authority in you: "*Thou art Goddess Mother, Thou art the Divine Child – Thou art all of That … a whole Universe,*" and bow deeply, as you recognize the Sacred before you.

Celebrant repeat the invocation/blessing to the first person sunwise, and pass them the bowl of oil.
Each may respond: It is so. I am She … (or however one wishes).

Cosmogenesis Dance
Celebrant: Let us celebrate this Sacredness, Her eternal Cosmogenesis, in the dance.

---

[387] This understanding of Air comes from Swimme, Ibid., 143-151.

REMIND ALL TO HOLD THE STILLPOINT WHEN IN CENTRE POSITION.

Start walk on 2nd measure so that all are linked on "Adoramus Te Domine" (if using that music): that is, start walk after 1st "Adoramus Te Domine."

MUSIC

The dance should finish with the group in the central linked position. Hold the final linked position for a while.

Sit in the Dark

Celebrant: "Let us sit in the Stillness." (quietly go to other location if ceremonial fire to be lit is elsewhere, perhaps inside).

LIGHTS OFF. CANDLES OUT. ALL MOVE INSIDE

Celebrant: "Let us sit in the stillness now, wrap the Dark Space of the Mother around."

One who will light fire has torch and matches ready.

Lighting of the Fire – after some time of sitting in the dark:

Celebrant announces: Out of Her fertile Dark Matter, out of the Stillness of her Creative Centre, New Being comes forth, Light is thus born. All Manifestation is born.

Co-celebrant moves to fire cauldron in centre, lights the fire, and announces: "We recall our Beginnings – the Great Flaring Forth, and Grandmother Supernova Tiamat, Goddess Mother of our Solar System – of our Star the Sun. This is our Cosmic lineage. We are Gift of Tiamat – Goddess Mother Supernova. Out of her stardust we are born. Carbon, hydrogen, oxygen, nitrogen, sulphur, phosphorus, and trace elements. We are Gift of Tiamat – out of her stardust we are born."[388]

Co-celebrant chants: "We are Gift of Tiamat – out of her stardust we are born."

---

[388] This is largely the composition of "The Tiamat Song" by Connie Barlow, *Green Space, Green Time* (New York: Springer-Verlag, 1997), 83.

Another repeats chant, then another repeats it.

Group joins in: "We are Gift of Tiamat – out of her stardust we are born." And continue for a few rounds, with the energy.

A participant (preferably a young one) lights a taper candle from the fire, and beckons:
"Let us emerge through the Gateway, coming into Being through the eons, taking our places in this moment."
When arriving at the altar, this person lights the old candle, announcing: "This is the old past year, and all it has brought," and then lights the new candle from this light: "and now we light the new." Blow out old candle.

Song
All sing: "Silent Night" cosmic version by Connie Barlow.[389] (Hand out torches and copies of song.)

Lighting the Candles
Celebrant: "The *Universe* wants to speak you, the Universe wants to speak *you*. Take a candle, light it, hold it up ceremoniously and recall, and speak if you like, of the new being coming forth in you this year."
Each in turn moves to the centre altar, takes a candle, puts a holder on it, lights the candle and speaks if they wish, then holds the lit candle up ceremoniously – which is the cue for all to respond with:

"So be it. May this new being within you come forth."

Song – when all have lit candles in hand:
Celebrant: "Let us celebrate these new beings coming forth, with the song."
All sing "PaGaian Joy to the World"[390] standing in a circle. (Hand out songsheets)

---

[389] See Appendix E. Also available at http://www.thegreatstory.org/songs/SilentNight.html.
[390] See Appendix F.

The Spiral Dance
Drop songsheets after singing complete song. All join hands each holding their candle in linked right hand (for moving sunwise to start). Celebrant leads into a spiral, all singing the first verse and chorus of "PaGaian Joy to the World" over and over, making sure to look in the eyes and faces of the passing people, then re-forming the circle.

Re-Generating the World
Celebrant: "Let us take our new beings – and re-generate the world, as the Sun has always done. What do you wish for the world, what is the flame in your heart? Join it with all the others."

Each one in turn steps into the centre and speaks if they wish, takes the holder off the candle, and puts their lit candle down firmly in the sand (best if the sand is wet) – which is the cue for all to respond with:

"May we be like the Sun and re-generate the world."

After all have put their candles down:
Celebrant: "May we re-generate the world with these new beings, as the Sun has always done."
Switch lights back on – wait while this is done.

Communion
Celebrant: "All glory and praise be to Her who daily gives us more than we can ask or imagine. We will now enjoy some of Her delights (hold up cake and wine), and remember that we also *are* Her delights.
You are Cake, for the Queen of Heaven and Earth, for the Universe. May you enjoy and be enjoyed.
You are Wine poured out for the Mystery, may your flavour be full."

Celebrant put wine down. Serve cake sunwise around circle, with the blessing. Each person in circle serve cake to next with blessing:
"You are cake for the Queen of Heaven and Earth – for the Universe. May you enjoy and be enjoyed."

Assistant take glasses to each person, AFTER all have cake.

Celebrant serve wine/juice to person in sunwise direction with blessing.
Each person in circle serve wine to next with the blessing:
"You are wine poured out for the Mystery. May your flavour be full."

Toasts – short OR Storytelling Circle (with ceremonial manners)[391]

Dance
Celebrant: "May we choose a joyful response to the awesome fact of our being in the Universe, and express that response through the art and dance of our lives.[392] You are the choreographer of your dance of life, we are the choreographers of the dance ... so enjoy!"

MUSIC:
I often use "Marco Polo" in Loreena McKennitt's *Book of Secrets* CD, but there are lots of possibilities.
Co-celebrants lead in improvised free dance, others may join in.

Open the Circle
All turning to the South
Celebrant: We have remembered this evening that we are Air – cosmic dynamics of exuberance – unfurling our being. May there be peace within us.
All: May there be peace within us.

All turning to the West
Celebrant: We have remembered this evening that we are Earth – deep memory and presence – remembering who we are. May there be peace within us.
All: May there be peace within us.

---

[391] These manners require listening with no comment or additions to the speaker's story, and confidentiality.
[392] This is a slightly altered quote from Matthew Fox. I can't remember where he said it.

All turning to the North
Celebrant: We have remembered this evening that we are Fire – vessels of the unseen shaping power of the cosmos – dancing. May there be peace within us.
All: May there be peace within us.

All turning to the East
Celebrant: We have remembered this evening that we are Water – cosmic dynamics of sensitivity – feeling what we are. May there be peace within us.
All: May there be peace within us

All turning to the Centre
Celebrant: We have remembered that we are Sacred Creativity, Goddess-Mother, Divine Child, all of That – that we are each and all Her Sacred Cosmogenesis. We have remembered our Origins, our Cosmic lineage and the new beings coming forth in us. We have remembered the flame in our hearts, and how we might re-generate the world, as the Sun has always done. May there be peace within us and between us. (all join hands)
All: May there be peace within us and between us.

Celebrant: May the Peace and Joy and Generosity of the Mother grow in our hearts and minds.
All: The circle is open, but unbroken
It has been a merry meeting,
It is a merry parting,
May we merry meet again.
Blessed Be.

MUSIC: I have used "Gregorian Waves" by Pascal Languirand (Imagine: 1991), but there are many possibilities.

Points for individual contemplation prior to the ceremony:
- the Centering/Breath meditation – all these birthings
- how you are Water, Fire, Earth and Air – Cosmic Dynamics

- how you and other are Goddess-Mother, Divine Child, Creator and Created … all That – a whole Universe.
- What new being is birthing in you this year.
- What do you wish for the world, how might you re-generate the world?

Winter Solstice/Yule Goddess Slideshow:
https://thegirlgod.com/pagaianresources.php

CHAPTER 6
IMBOLC/EARLY SPRING

Southern Hemisphere – August 1st/2nd
Northern Hemisphere – February 1st/2nd
These dates are traditional, though the actual astronomical date varies. It is the meridian point or cross-quarter day between Winter Solstice and Spring Equinox, thus actually a little later in early August for S.H., and early February for N.H., respectively.

In this cosmology Imbolc/Early Spring is the quintessential celebration of *She Who is the Urge to Be*. This aspect of the Creative Triplicity is associated with the *differentiation* quality of Cosmogenesis,[393] and with the Virgin/Young One aspect of the Triple Goddess, who is ever-new, unique, and singular in Her beauty – as each being is. This Seasonal Moment celebrates an *identification* with the Virgin/Young One – the rest of the light part of the cycle celebrates Her *processes*. At this Moment She is the Promise of Life, a spiritual warrior, determined to Be. Her purity is Her singularity of purpose. Her inviolability is Her determination to be … nothing to do with unbroken hymens of the dualistic and patriarchal mind. The Virgin quality is the essential "yes" to Being – not the "no" She was turned into.

In the poietic process of the Seasonal Moments of Samhain/Deep Autumn, Winter Solstice and Imbolc/Early Spring, one may get a sense of these three in a movement towards manifest form – syntropy: from the *autopoietic* fertile sentient space of Samhain, through the gateway and *communion* of Winter Solstice to *differentiated* being, constant novelty, infinite particularity of Imbolc/Early Spring. The three are a kaleidoscope, seamlessly connected. The ceremonial breath meditations for all three of these Seasonal Moments focus attention on the Space between the breaths – each with slightly different emphasis: it is from this manifesting Space that form/manifestation arises. If one may observe Sun's position on the horizon as She rises, the connection of the three can be noted there also: that is, Sun at Samhain/Deep Autumn and Imbolc/Early Spring rises at the same position, halfway between Winter

---

393 Swimme and Berry, *The Universe Story*, 73-75.

Solstice and Equinox, but the movement is just different in direction.[394] And these three Seasonal Moments are not clearly distinguishable – they are "fuzzy,"[395] not simply linear and all three are in each other … this is something recognised of Old, thus the Nine Muses, or the numinosity of any multiple of three.

Some Imbolc/early Spring Story

This is the Season of the new waxing light. Earth's tilt has begun taking us in this region back towards the Sun. Traditionally this Seasonal Point has been a time of nurturing the new life that is beginning to show itself – around us in flora and fauna, and within. It is a time of committing one's self to the new life and to inspiration – in the garden, in the soul, and in the Cosmos. We may celebrate the new young Cosmos – that time in our Cosmic story when She was only a billion years old and galaxies were forming, as well as the new that is ever coming forth. This first Seasonal transition of the light part of the cycle has been named "Imbolc" – Imbolc is thought to mean "ewe's milk" from the word "Oimelc," as it is the time when lambs were/are born, and milk was in plentiful supply. It is also known as "the Feast of Brigid," Brigid being the Great Goddess of the Celtic (and likely pre-Celtic) peoples, who in Christian times was made into a saint. The Great Goddess Brigid is classically associated with early Spring since the earliest of times, but her symbology has evolved with the changing eras – sea, grain, cow. In our times we could associate Her also with the Milky Way, our own galaxy that nurtures our life – Brigid's jurisdiction has been extended.

Some sources say that Imbolc means "in the belly of the Mother." In either case of its meaning, this celebration is in direct relation to, and an extension of, the Winter Solstice – when the Birth of all is

---

[394] It is a very informative and simple process to note this in your place, to make markers of some kind, and observe it over the period of the year.

[395] "Fuzziness" is a term used by scientist and philosopher Vladimir Dimitrov, who describes that according to fuzzy set theory, the meaning of words cannot be precisely defined. See Dimitrov, *Introduction to Fuzziology*. Also here: http://www.zulenet.com/vladimirdimitrov/pages/fuzzycomplex.html.

celebrated. Imbolc may be a dwelling upon the "originating power," and that it is in us: a celebration of each being's particular participation in this power that permeates the Universe, and is present in the condition of every moment.[396]

This Seasonal Moment focuses on the *Urge to Be*, the One/Energy deeply resolute about Being. She is in that way – and Self-centred. In the ancient Celtic tradition Great Goddess Bri wilful gid has been identified with the role of tending the Flame of Being, and with the Flame itself. Brigid has been described as: "… Great Moon Mother, patroness (sic … why not "matron") of poetry and of all 'making' and of the arts of healing."[397] Brigid's name means "the Great or Sublime One," from the root *brig*, "power, strength, vigor, force, efficiency, substance, essence, and meaning."[398] She is poet, physician/healer, smith-artisan: qualities that resonate with the virgin-mother-crone but are not chronologically or biologically bound – thus are clearly ever present Creative Dynamic. Brigid's priestesses in Kildare tended a flame, which was extinguished by Papal edict in 1100 C.E., and was re-lit in 1998 C.E.. In the Christian era, these Early Spring/Imbolc celebrations of the Virgin quality, the New Young One – became "Candlemas," a time for purifying the "polluted" mother – forty days after Solstice birthing. Many nuns took their vows of celibacy at this time, invoking the asexual virgin bride.[399] This is in contrast to its original meaning, and a great example of what happened to this Earth-based tradition in the period of colonization of indigenous peoples.

The flame of being within is to be protected and nurtured: the new Being requires dedication and attention. At this early stage of its advent, there is nothing certain about its staying power and growth: there may be uncertainties of various kinds. So there is traditionally a "dedication" in the ceremonies, which may be considered a "Brigid-ine" dedication, or known as a "Bridal" dedication, since "Bride" is a

---

396 Swimme and Berry, *The Universe Story,* 17.

397 Durdin-Robertson, *The Year of the Goddess*, 36, quoting Denning M. et al, *The Magical Philosophy III* (Minnesota: St. Paul, 1974), 166.

398 Dames, *Ireland*, 233.

399 See Crowley, *Celtic Wisdom,* 57.

derivative of the name "Brigid."[400] A commitment to Brigid, to Her Virgin integrity (as it has been defined earlier), is a dedication to tending your-self, understanding that self (Self) as an expression of the Whole: any being – female, male and all gender/sexes – may so dedicate themselves. Imbolc celebrations may a kind of wedding, of the Beloved and the Lover, the small self with Deep Self: it may be understood as the original wedding – as the mystics say. It is the Beloved and Lover in awe of each other – "in Love": that is, I am the Mother's (who is the Beloved) before I am anyone else's, I belong to Her first and foremost (as each being does). This is the quintessential wedding. Yet it is also maternal – a dyadic unity[401] ... one may image it this way. Thus at Imbolc, there may be mixed metaphor, of nurturing the new unique being, and also being in Love.[402] Perhaps it is that the maternal relationship at its best, and the kind of thought that it requires, may be a model for all other relationships.

Also of note is that the word "brigand" has its roots in Brigid's name, and it came to mean "outlaw," and it may be asked: "outside of whose law?" So, a devotee of Brigid may be a *brigand*, following an Earth lore perhaps. This then is a possible expression of the Brigid-ine dedication in the ceremony. Imbolc/Early Spring is the time for strengthening the *Brigid* in you, the *brigand* or the *Bride*, however one may like to express and understand Her power within. It is also possible to extend the maternal, pastoral and/or agricultural metaphor to "husbanding" in the original and true sense of that word, as in a caretaker: a "husband" may be understood as a "Mummy's boy," a devotee of the Mother.

Brigid may be praised as of "the holy well and the sacred flame."[403] The "well" designates a source of life, to which the new tender being is intimately connected. It may be understood as the Well of

---

400 Is this why so many women want to be Bride – even just for a day, even if it may mean a lifetime of dedication to another?

401 A term used by Gross, "The Feminine Principle in Tibetan Vajrayana Buddhism."

402 Imbolc is a good Moment to quote mystical poets such as Jami and Iraqi, changing the sex of the metaphor mostly used.

403 Starhawk, *Truth or Dare*, 289-295.

Creativity of the Cosmos[404] to which all are seamlessly connected. A dedication to Brigid means a dedication to the being and beauty of particular small self, and knowing deeply its Source, rooted seamlessly in the whole of Gaia – as the infant knows deeply its dependence on the Mother, as the new shoot on the tree knows intimately its dependence on the branch and whole tree, as the new star's being is connected to the supernova. The "Well" of Brigid then, is not only the well into Earth-Gaia, but also Universe-Gaia's Well of Creativity. Brigid has extended Her "country," Her jurisdiction; and perhaps it was always understood that way by the ancients.

Imbolc ceremony may be an invocation of the sheer magic of the *energy to be.* If one has ever been deprived of it – in depression, illness, uncertainty, the mists of apathy – one knows the beauty and preciousness of this *Urge to Be*: and it may be consciously nurtured, tended, and rejoiced in – Life/Creativity WILL proceed! It is a blessed thing – it is an Annunciation; in this Cosmology we all bear the Promised One. We *are* the Promised One. Each has a particular Creativity to deliver that no-one else can, and this ceremonial moment is an opportunity to say "Yes" and commit one's self to the flourishing of your small part, which is a totally unique beauty in the history of the Universe. Mary's "Fiat" in the Christian tradition can be seen this way – but unfortunately it is used to support a dominating, colonizing power structure. In the PaGaian Imbolc ceremony, Mary's yes is reclaimed in the context of saying "Yes" to each one's particular Creativity, each one's responsibility as a *Promise of Life.*

Artemis is another Goddess who has been traditionally understood to midwife new being, and may be invoked at this time; herbs sometimes used to ease the pain of childbirth were named *Artemisia.* Artemis is also an archer, the one whose arrow flies true and on centre – Imbolc is a good Seasonal Moment to remember the power and presence of the desire to fly true and on centre. Coming into being takes courage and daring: thus the Virgin aspect is often understood as spiritual warrior. Esther Harding has described the Virgin quality thus:

---

404 See Swimme, *Canticle to the Cosmos*, video 10 for more on Creativity.

> The (one) who is virgin, one-in-herself, does what she does – not because of any desire to please, not to be liked, or to be approved, even by herself; not because of any desire to gain power over another ... but because what she does is true.[405]

The 'Virginal' motivation for action is not guilt, fear, envy … but the truth of one's being. This is Her purity and her freedom, and what we may seek to affirm in ourselves. Traditionally Imbolc is a time of purification and strengthening;[406] that is what the nurturing of the new requires – care and vigilance and sometimes discipline.

There is risk and resistance to coming into being. The Universe itself knew it when it encountered gravitational resistance to expansion, in our very beginnings, in the primordial Flaring Forth. It was never without creative tension. The Universe knows it daily, in every moment. Imbolc can be a time of remembering personal vulnerabilities, feeling them and accepting them, but remaining resolute in birthing and tending of the new, listening for the Urge of the Creative Universe within. Brian Swimme has said (quoting cultural anthropologist A.L. Kroeber) that the destiny of the human is not "bovine placidity" but the highest degree of tension that can be creatively born.[407]

For women in the patriarchal cultural context, the Imbolc process/ceremony may be an important integrating expression, used as they might be, to fragmentation in relationship – giving themselves away too easily. This seasonal celebration of individuation/differentiation, yet with integrity/wholeness, especially invoking She-who-is-unto-Herself, can be a significant dedication. However, most Gaian beings have uncertainties, and many specifically about gender/sex and aspects of wholeness and integrity; all new being requires care and commitment. Men also need to return to relationship with the Mother, in themselves and in GaiaEarth: and may choose this Seasonal Moment to re-identify with the ancient traditions of the male as a Caretaker, Shepherd, Gardener – "at-tending," and Son of the Mother.

---

405 Harding, *Woman's Mysteries: Ancient and Modern*, 125.

406 Starhawk, *Truth or Dare*, 304.

407 *The Canticle to the Cosmos*, video 8, "The Nature of the Human."

Imbolc/Early Spring Breath Meditation

To begin this Seasonal Moment's contemplation:

Centering for a moment … feeling the full cycle of your breath. Recognising the Triple Goddess Dynamic here in this primary Place: She who unfolds the Cosmos … receiving the Gift, as you inhale – feeling the Urge to Be, feeling the peaking – the exchange, the communion that this Place is, then the release – the dissolving, becoming the Gift, making space, ever transforming … the Triple Spiral dynamic here in your breath, the Wheel of the Year in your breath.

Some Imbolc/Early Spring motifs

[Figure 13]

Suggested decorations for this Seasonal Moment – on the front door, windowsills, corners of shelves, and/or altars, may include:

- **white** – representing the new growing light.
- **a tea candle** – representing the small self, and the flame of being.
- **light blue ribbon** – blue associated with young hot flame.
- **a mirror** – for self-knowledge and love of self.

- **a tiara** – representing primacy of self, Bride-al. It could be a crown/garland of some kind.
- **an arrow** – flying straight and true, Artemis association.
- **photos of the young one, and the innocent** – pre-informed, pre-domesticated, the wild.
- **Artemis** herbs – easing the pains of coming into being.
- **delicate white lacy fabric** – representing the new web of being, the delicate web of life. The early Universe itself, the protogalactic clouds, had a lacy pattern caused by quantum fluctuations that enabled the galaxies to form. Swimme and Berry describe the young billion year old universe:
  After the fireball ended, the Universe's primordial blaze extinguished itself, only to burn once more in the form of lacy veils of galaxies filling all one billion years of space-time existence.[408]
- **poetry** – Brigid's Spelling, self-expression.

An Invocation
VIRGIN[409]
Virgin,
– warrior of the spirit,
urging me on
calling forth my green shoots
singing them out of the darkness.
in love with this new life
no-one can distract you.
Yet you do not curse the darkness
you know it and trust it.

After the storm, the tempest, the destruction,
in the compost,
Is there a will to live?
something to grow?
From whence will come the mighty forest?
Where does it begin?

---

[408] Swimme and Berry, *The Universe Story*, 34.
[409] A poem by Glenys Livingstone, 1995.

Artemis, great Virgin Goddess
of the hibernating bear who wakes in the Spring
of the deer whose antlers regrow so fast

Protect our spirits, our will to life
tend the Flame within us.

The Universe Story and Imbolc/Early Spring

Cosmologist Brian Swimme has described the human as a "perpetually young species." He said:

> What life did when it created the human was to take infancy, stretch it out, and call it a species. We are constantly within the creative movement and exploration of a mammalian infant. This is called neotyny. The deep dynamic of the child is at the core of biological creativity.[410]

He went on to admonish that the mind of the child is a goal of all education.

In *The Universe is a Green Dragon*, Brian Swimme describes "the source of all that is, the support and well of being" as "Ultimate Generosity," noting how elementary particles erupt out of no-thing-ness: that the ultimate realm of generation, the *Creative Space*, is emptiness, because "Every *thing* ... all existence has been poured forth,"[411] IS poured forth. The Universe is one huge self-expression of generosity, and hence that "self-expression is the primary sacrament of the Universe."[412] He describes: "We are Generosity-of-Being evolved into human form."[413]

Imbolc celebration may be understood to be a Moment for dedication to *Original* relationship with Place of Being, this Universe/Cosmos, one's *original* nature, the pre-informed, the wild: one's Gaian indigeneity. We – as a species and as individuals – are in fact, in "the belly of the Mother," always have been, and always as new beings,

---

410 *The Canticle to the Cosmos*, video 8, "The Nature of the Human."
411 146.
412 Ibid., 147.
413 Ibid.

and even yet to understand ourselves. Each particular being is a seamless continuation and expression of the *Original Ovulation.*

At Imbolc/Early Spring, the continued birthing – the rushing away from Origins, the continued rippling forth of Creation – may be celebrated with the understanding of the difficulties, the resistances that even Gaia-Universe has encountered, and how this has served the unfolding of the story as we know it. Imbolc may celebrate Gaia's rush to diversity, differentiation; we commit ourselves to this, beginning with ourselves. The ceremonial process of "purification and strengthening" may be understood as a feeling for where it is in us that the Universe is acting now – where each one feels the excitement of Creativity calling to them in their lives.[414]

And a quote from *The Universe Story*, that is resonant with Imbolc/Early Spring:

> ... the first living cell .... emerged from the cybernetic storms of the primeval oceans... Life here was born in a lightning flash. ...(the first cells – prokaryotes) were the most fragile autopoietic structures yet to appear ... and yet they were essential for the next advance... For four billion years the prokaryotic organisms have been remembering the composition from the beginning. ... Though fragile, though liable to destruction and change in an infinity of ways, they could nevertheless perform an aboriginal magic that would enable them to pervade the world: they could swallow a drop of seawater and spit out a living version of themselves. ... Besides these new powers of autopoiesis, cells exhibited a new depth of differentiation as well. Once every million births, a cell was created that was new.[415]

Notes on imagery and metaphors in the offered ceremonial script

The Centering Breath Meditation begins with the ringing of a small bell enhanced with the wooden stick, as it may suggest the solitary tenuous beauty of the small self, the Cosmogenetic quality of

---

414 Swimme speaks of such feeling in *Canticle to the Cosmos*, video 10 "The Timing of Creativity."

415 Swimme and Berry, *The Universe Story,* 86-88.

differentiation, each being's unique single small self that is celebrated particularly in this beginning of the light part of the cycle.

The focus on the breath begins with the Space between the breaths, as it did for Samhain and Winter Solstice, with the emptiness; but now imagining it as the "Sea of Generosity" from which all springs forth. The experience of the breath then is imagined as a "ripple stirring upon the Sea," extending to imagining the self as being that ripple. This is inspired by the mystical poetry of Iraqi, which is spoken, where he expresses that on "the morning of manifestation … ripples stirred upon the "Sea of Generosity."[416]

The breath is then used to speak "the Mother's word of Creation – Om," contemplating how we "spell" the world personally and collectively. We may understand that all of manifestation is the exuberant expression of Great Goddess/Deity/Mystery, that all of manifestation is Her Poetry, as many poets and mystical traditions have. The letters of the Sanskrit alphabet were called "mothers," matrika. "*Om* was the *mantramatrika*, Mother of Mantras; and these divine Words spoken by Kali created and destroyed everything, including all gods."[417] It has derivatives in many cultures. In many cultural creation stories the world proceeds from sound: for some it is cosmic music that is Creatrix, as in the case of Magoist Cosmogony of East Asia.[418] Modern Western science has discovered that sound ripples out of black holes.[419] The essence of speaking the word then in the Imbolc ceremony, is the *spelling* – how we co-create the world with our expression. Imbolc is a time for being mindful of our creations: we join and weave our Om's together … we do co-create.

For Calling the Directions/Elements, Casting the Circle, the focus is on the internal physical experience of the elements within each particular self: that is, on the consciousness of the space within our organism, and how we can sense/feel the elements in our manifest form,

---

[416] Chittick and Wilson, trans., *Fakhruddin 'Iraqi: Divine Flashes.* See https://en.wikiquote.org/wiki/Fakhruddin_%27Iraqi.
[417] Walker, *The Woman's Encyclopedia of Myths and Secrets*, 546.
[418] Hwang, *The Mago Way*, 127.
[419] NASA report 2003: https://science.nasa.gov/science-news/science-at-nasa/2003/09sep_blackholesounds.

our bodyminds. At Imbolc then, the external actual elements around the altar are not actually handled as usual, but attention is simply paid to the personal internal sensation of each.

For the Invocation at this Seasonal Moment when the flame/light is tender and new, and when it is the "Urge to Be" that is celebrated and nurtured, participants each light a candle from the centre candle and gaze into the flame, contemplating how the Flame of Being is within them. In their own time each one places the candle on the altar affirming in their own words the presence of this Virgin/Young One quality in their being. The midwifing and "true" qualities of Artemis are then invoked with the tasting of Artemisia herbs and the holding of Her arrow. Sometimes this "Artemis process" has been done later in the ceremony, before Opening the Circle.

The Communion food and drink at Imbolc, this first celebration of the light part of the cycle, are light in colour or white. The words of blessing are an adaptation of the words traditionally attributed to the angel's words to Mary in Christian tradition – the "Annunciation" as it is known. This is a reclaiming of that metaphor of Annunciation for all; that is, as a recognition of the Holy One, as the Young Promised One, present and growing in the self.

In the offered script there is a red aisle for procession for the rite of the Brigid-ine dedication – for the 'Bridal'/'brigand' process. This process has its roots in the understanding that the red aisle represents the blood of the Mother, "the wise blood," and that is integral to Creativity. To walk on the red processional aisle is to be honored and recognized as participating in the Wisdom of the Mother, to be in Her Creative flow and timing: to be held and safeguarded within Her creative flow.

When Brigid is invoked for this rite of dedication, the singing of "Ave Brigid" by the group is in the style of the Christian "Ave, Ave Maria,"[420] reclaiming this as praise of Great Goddess.

A most enjoyed part of the Imbolc ceremony is the circle dance done after all the dedications, with the invitation to dance our commitments "into Being." It is known as the "Misirlou Circle Dance," as I learned it originally from Jean Houston by that name.[421] It recalls

---

[420] https://thegirlgod.com/pagaianresources.php.

[421] It may be seen here: https://thegirlgod.com/pagaianresources.php

ancient traditions, how we may have danced in the temples, uniting our hearts and minds, with the well of Creativity.

[Figure 14] An Imbolc/Early Spring Ceremonial Altar

AN IMBOLC/EARLY SPRING CEREMONIAL SCRIPT

Note that the directions here are called in "counter-clockwise" direction which is sunwise in the Southern Hemisphere. The order may be changed to "clockwise"/sunwise for the Northern Hemisphere. Note that this particular offered ceremonial script is not considered prescriptive, and that there are many creative possible variations.

Some requirements and some suggestions: Large "well" of salt water with clay bowl of sand for flame held within it – on altar (see above photo

---

And for dance instructions and a link to music: https://pagaian.org/articles/misirlou-dance-music-and-instructions/.

of Imbolc ceremonial altar). White/creamy and lace altar cloth. Smaller bowl of salt water, hand towel. Each wear 'Brid-al'/'brigand' gear, and a 'crown'/veil. Arrow on altar, bow somewhere if possible. Fruit bread, fruit juice already poured in glasses on tray, white chocolate, ricotta, feta cheese. Seasonal fresh greenery, buds. Bouquet of Artemisia herbs tied with silver ribbon for each. Basket of taper/dinner candles & trays of damp sand. Red cloth/carpet laid down. Misirlou music for dance. Copies of 'Bridal'/Brigidine commitment. Sprigs of seasonal flower (wattle in Australia) in vase near red carpet. Warm bell and stick. Three firepots around edge of circle (and optional fireplace) ready to light. Images of invoked Goddesses visible in circle.[422]

Call to Gather Drumming – in honour of Great Goddesses Brigid and Artemis, who call us into Being.

Centering – Breath Meditation "Listen to the bell as if listening to the sound of your unique Beauty ringing – your singular particular Beauty, unique in the history of the Universe, ringing." Ring the bell, accentuate with the stick and continue for a few minutes.

"Take a deep breath, and as you let the breath go, follow it down to the Space before your next breath. Imagine this Space, this Emptiness, as ***the*** Sea of Generosity … from which All springs forth. Feel the Urge of Creation springing forth – as the Urge to Breathe arises. Imagine this breath as a Ripple upon the Sea of Generosity. Feel it rippling out of this Sea as your breath expands. Imagine your-Self as this Ripple. We are ripples stirring upon the Original Sea of Generosity now."

The Morning of Manifestation sighed,
the breeze of Grace breathed gently,
ripples stirred
upon the sea of Generosity.
The clouds of Abundance poured down the rain
upon the soil of preparedness;

---

422 See Imbolc Goddess Slideshow link at the end of the chapter for the images named in this ceremonial script.

so much rain that the earth shone with Light.[423]

Speaking the Word of Creation

Next time you take a deep breath, we will do it together, and as we let the breath go, let us speak the Mother's word of Creation "Om," the MantraMatrika – or some variation of it – let it come from your belly. You may contemplate how you spell the world, co-create it with your self-expression, and how it weaves with others. Let us Spell it together, let us Conspire.

Statement of Purpose

This is the season of the waxing light. Earth's tilt is taking us back towards the Sun. The seed of light born at the Winter Solstice begins to manifest, and we who were midwives to this infant flame now see it grow strong as the light part of the day grows longer. This is the time when we celebrate individuation: how we each become uniquely ourselves. It is the time when we celebrate beginnings … the first tendrils of green emerging tentatively from the seed. We meet to share the light of inspiration and creative intentions, which will grow with the growing year.

This is the Feast of the Virgin – Brigid, She who tends the Flame of Being; Artemis, She who midwifes body and soul. She is the *Urge to Be*, deeply committed to the Creative Urge, to manifestation, deeply committed to Self and to bringing forth the New, the Promised One. She is uncompromised, unswerving, noble, true, a warrior of spirit. She will protect the stirrings of Life.

Calling the Directions – Casting the Circle

Let us begin our celebration of Her, by remembering our beginnings, from whence we come – our ***true*** nature.

Celebrant go to the East – all may turn. DRUM
"We are from the East, and we are Water – filled with the primordial oceans. You may feel Her in you, Her moistness, taste it. We are juicy with it, soft with it. Take a moment to feel how you are Water. (pause) We are Water, we are this."

---

423 From *Fakhruddin 'Iraqi: Divine Flashes,* trans. Chittick and Wilson.

Response: We are Water, we are this.

Celebrant go to the North – all may turn. DRUM
"We are from the North, and we are Fire – sparks of ancient heat. You may feel Her in you, the warmth of your body. Your metabolism an echo of the Great Fire at the Beginning, and of our Sun. Take a moment to feel how you are Fire. (pause) We are Fire, we are this."
Response: We are Fire, we are this.

Celebrant go to the West – all may turn. DRUM
"We are from the West, and we are Earth – geological, Gaian formations. You may feel Her in you – the weight of your body, your memories. She is alive in us and we in Her. Take a moment to feel how you are Earth. (pause) We are Earth, we are this."
Response: We are Earth, we are this.

Celebrant go to the South – all may turn. DRUM
"We come from the South, and we are Air – drawn from and drawing from this ancient river that all have breathed. You may feel Her in you, expand with Her. We are inspired with Her. Take a moment to feel how you are Air. (pause) We are Air, we are this."
Response: We are Air, we are this.

Celebrant: This is our true nature – Water, Fire, Earth, Air (PACING THE CIRCLE perhaps). We have always been from the very beginning, and so we always shall be. We are at the Centre of Creativity – a multiplicity of centres. The circle is cast, we are between the worlds, beyond the bounds of time and space, where light and dark, birth and death, joy and sorrow, meet as One.
(Celebrant light centre candle)

Invocation
Pass basket of candles, each take one.
Co-celebrant light a dinner candle from the centre flame and take the flame to each one, with blessing:
"Receive the Flame from the Mothers – who have birthed you and all."

OR each may come forward and light their candle from the centre candle.

When all have their candle lit, Celebrant/Co-celebrant:
"This flame recalls the Great flaring forth:
the primordial beginnings, the flame of being, the originating power."[424]

Response: "It is so" or … (as one wishes)

Celebrant: "Sit for a few moments gazing into the flame. You may contemplate how this flame is in you, your vital lifeforce, your warmth and heat. You are a spark of the beginning, an individual manifestation of the becoming universe"

SITTING QUIETLY FOR A FEW MOMENTS.
"When you are ready, one at a time, place your flame on the altar, speak if you wish, affirming the gift of this Flame, this Original Flame in you, this spark from the beginning, the ever-new young one that you are … however you'd like to express it, if you would."

Each put candle down in the damp sand when ready to.
Each one: "I am this Flame … (some expression as is meaningful).
AND/OR offered response: "The Original Flame of Being, the Originating Power is in me. I receive it, and in me it is ever-new."

After all the candles are on the altar:
Celebrant: "We are small flames of Being … a multiplicity of centres."

Artemis Process
Celebrant hold up the herbs and arrow and pronounce: "Let us invoke Artemis, She of Old who brings forth the New, whose herbs ease the pain of birth, and whose arrow flies true."

Pass the basket of Artemisia, each take a bouquet.

---

[424] I acknowledge Adriana Zapata-Delgado and Zoe Anderson (2016) in the some of the composition of this script.

Celebrant hold up herbs: "Taste her herbs, to remind you of the creative resistance of coming into Being, of birthing the New – may She midwife you."
Together each rub and taste the bitterness.
Short response optional: "May it be so." Or "I remember …"

Celebrant holds up the arrow:
"And hold Her arrow, to remind you to be true."
Pass arrow around circle for each to hold to their heart.
Short response optional: "May I be true, or …"
After all are done:

She is faithful to the heart that seeks
fullness, wholeness
She will constellate in the heart's night sky
and call forth right direction.[425]

Purification and Strengthening[426]
Celebrant: "Let us now begin our invocation of Brigid.
Take the bowl of salt water, recall, wash, and speak briefly if you wish, of what weakens the Flame within you, what leaches your power?"
Pass bowl of salt water, counter-sunwise
Each one wash, and may speak briefly as they wish, or a word or phrase.
Response: May you be cleansed.

When the bowl has gone full circle, and all have washed:
Celebrant: "Let us raise some power, some energy."

DRUM begins, and each may begin to move/dance, improvised voice continues, percussion, and moving as each wishes.

---

[425] From *Virgin*, poem by Glenys Livingstone, 1995.
[426] This process with the salt water is adapted from Starhawk, *Truth or Dare,* 304-305.

(Suggest the pose of Eurynome,[427] the Bird-Headed Snake Goddess: that is, with arms raised?)

When energy is raised, return to ground, to sitting.
Celebrant: "Wash in the water again, and speak briefly if you wish, of what strengthens the Flame, the Urge to Be – the power within you."
Pass bowl of salt water, sunwise.
Each one wash, and may speak briefly as they wish, or a word or phrase.
Response: We bless your power.

When the bowl has gone full circle, celebrant pour water into the centre bowl, with blessing:
"We commend our power to and from its Source – the Well of Creativity within, and in Whom we are immersed."
Co-celebrant light the centre flame (which is in the 'Well' on the altar) and announce: "We are tenders of the flame. May we be Brigids, Brides, Brigands – tenders of the Flame."

CO-CELEBRANTS LIGHT OTHER FLAMES AROUND THE CIRCLE – all singing:
"We are tenders of the Flame" – or other Brigid song.
Celebrant: "Take up your flame now and let us invoke the ancient One Brigid, She within who tends the flame of Being."

Each one takes up their candle, standing.
All sing: "Ave, Ave, Ave Brigid" X 2 as in the "Ave, Ave, Ave Maria" until Brigid appears.

Celebrant (becoming Brigid) – mask on, take up bundle of seasonal flower, walk the circle, and stand near red carpet.

Co-celebrants exclaim: "Brigid!"

---

[427] See Austen, *The Heart of the Goddess*, 8, and also the Goddess Imbolc Slideshow linked at the end of the chapter.

The "Bridal"/Brigid-ine/brigand Dedication

Celebrant (Brigid): "Now is the time for your commitment to Being, your dedication to Self. Step forward one at a time, process up and down the aisle, in Her creative flow: then pause and speak if you wish, of your dedication to self and to being."

Each step forward in turn, process up and down the aisle once or a few times (holding up their candle, arms raised if they wish).
Each stop facing the circle, and speak if and as they wish:
(Create own words, or option to speak these or a variation)
"I commit myself to my particular small self, understanding that I am She – Gaia – She who is All. I am connected to Her as the tree bud is to the branch. 'I am the beauty of the green earth and the white moon among the stars and the mystery of the waters'. I commit myself to this Originating Power present in me, the Sacred Flame in me, this Native Land who I am. I will protect Her and honour Her in myself – this particular Beauty, who is ever-new. I am a Promise of Life. Whatever She needs I will give Her. I will tend Her in myself – so that She may grow strong and flourish."

Celebrant (Brigid): "Rejoice O highly favoured. Gaia Herself rejoices in your commitment and freedom to Be, for you are a Promise – of Life."

Brigid gives each a sprig of seasonal flower.

Group pronounces: "Ave (person's name)!"

Celebrant takes off Her mask (and headpiece) and unwinds (walking circle in a counter-sunwise direction): a co-celebrant takes on Her mask, walks the circle sunwise, then plays Brigid's role for the celebrant. When done, she takes off Brigid's mask and "unwinds."

After all have made commitments and processed:

Celebrant: "Let us call upon the Powers in us":

Celebrant gesture to Goddess images[428] in turn:
"Brigid of the holy well and the sacred flame!"[429]
Group choruses: Brigid of the holy well and the sacred flame!

Celebrant: "Artemis of the arrow, flying true and on Centre!"
Group choruses: Artemis of the arrow, flying true and on Centre!

Celebrant: "Eurynome, Ancient One, whose passion is Life!"
Group choruses: Eurynome, Ancient One, whose passion is Life!

Celebrant: "Aphrodite, who sings love songs to her own beauty."
Group choruses: Aphrodite, who sings love songs to her own beauty.

Celebrant: "The Lover, Radha, who sees Who She really is."
Group choruses: The Lover, Radha, who sees Who She really is.

Celebrant: "We invoke you and celebrate you in ...." (group names each around the circle).

Dance
"Let us celebrate our commitment to Being in the dance. Let us dance it into Being." All put candles in sandtrays on altar.
Co-celebrant re-light centre flame.
MUSIC ON (Misirlou music)

Communion
Celebrant holding up fruit bread, and fruit juice:
"Let us further enjoy our manifested state with food of the Mother, brought to us by Her and by the creativity of our ancestors."

Server holding tray of poured juice.
Celebrant offers glass of juice and fruit bread to each, with the blessing/annunciation:
"Rejoice O highly favoured

---

428 See the end of the chapter for the particular images named here.
429 Starhawk, *Truth or Dare*, 289.

Blessed are you, and blessed is the fruit of your Creativity."

Response: "Fiat!" or "May it be so." (and/or whatever else as one wishes)

Pass around white chocolates, ricotta, feta cheese, white wine.

Story Space (with ceremonial manners)[430]
Share stories of creative projects and intentions.

Open Circle
We have remembered this evening that we are Air, we feel Her in us – this ancient river, expanding ... may there be peace within us.
All: May there be peace within us.

We have remembered this evening that we are Earth, we feel Her in us – our weight, a geological formation ... may there be peace within us.
All: May there be peace within us.

We have remembered this evening that we are Fire, we feel Her in us – our warmth an echo of the Great Fire... may there be peace within us.
All: May there be peace within us.

We have remembered this evening that we are Water, we feel her in us – our moistness, the primordial ocean ... may there be peace within us.
All: May there be peace within us.

We have remembered that we are creative Ripples stirring upon the Original Sea of Generosity, ever-new Flames of Being, spelling the world with ourselves. We have dedicated ourselves to tending the Flame of Being – this Brigid-ine One, this New Young One, the Beloved of self – She who is All. We have celebrated these dedications in the Dance. May there be peace within us & between us. (all join hands)
All: May there be peace within us & between us.

---

[430] These manners require listening with no comment or additions to the speaker's story, and confidentiality.

Celebrant: May the Truth and Power of ever-Virgin Goddess be in our hearts and minds.

Song: The circle is open but unbroken, may the peace of the Goddess be ever in our hearts, merry meet and merry part and merry meet again. Blessed Be!

Points for individual contemplation prior to the ritual:

- how the Flame of Being is given to you, is within you, and is ever-new in you: this Original seamless relationship, and your particular unique self, this Virgin-self, pre-informed, wild, native.
- What weakens this Flame within you – your strength of Being?
- What strengthens this Flame of Being within you? What strengthens the life-force within you?
- Your commitment/dedication to self – how you might express that … with the offered words and/or your own? or even simply the statement of standing there with candle flame?

Imbolc/Early Spring Goddess Slideshow:
https://thegirlgod.com/pagaianresources.php

Images referred to in the Imbolc/Early Spring Ceremonial Script:

[Figure 15] Brigid (Photo credit: Glenys Livingstone, MoonCourt Imbolc altar.)

[Figure 16] Artemis (Image credit: Hallie Iglehart Austen, 53).

[Figure 17] Eurynome (Image credit: Hallie Iglehart Austen, 9).

[Figure 18] Aphrodite (Image credit: Hallie Iglehart Austen, 133).

[Figure 19] Radha (Image credit: Glenys Livingstone, personal statue).

CHAPTER 7
SPRING EQUINOX/EOSTAR

Southern Hemisphere – September 21-23
Northern Hemisphere – March 21-23

Spring Equinox/Eostar is the midpoint between Winter and Summer Solstices in the light part of the annual cycle; it is the point of balance in that part of the cycle. The young and growing light has come into balance with the dark. The Sun is equidistant between South and North on the horizon, yet the trend at this Equinox is toward increasing light and warmth: thus it may be understood poetically as a "stepping into the power of being," as life does at this time. In this cosmology it has been storied as the joyful return of the Beloved One who has gained new wisdom after long and lost wanderings: in particular the return of the Daughter Persephone from the Underworld, as in the tradition of the Lesser Eleusinian Mysteries of Greece, where for millennia Her return to Her Mother Demeter was celebrated in the Spring.

In PaGaian Cosmology, Persephone's descent and return is largely understood as voluntary, as Charlene Spretnak tells it in her book *Lost Goddesses of Early* Greece;[431] thus the story retains the organic nature of this essential process of being, and Persephone may truly be understood as metaphor for the Seed of Life, a consciously acting redeemer figure. As Seed of Life in this Moment of the annual cycle, She has navigated the darkness successfully, has gained wisdom from the depths, and will flower, and come to fruition. Other cultures have also celebrated Spring Equinox as the return of the Beloved One: for example, Dumuzi in Sumer. In the oldest stories of Persephone, She descends to the underworld voluntarily as a courageous seeker of wisdom, a compassionate receiver of the dead, a redeemer figure. She is the *Seed of Life* that never fades away. Spring Equinox is a celebration of Her return, life's continual return, and also our own personal emergences/returns. We may contemplate the **collective** emergence/ return also – especially in our times.

In the PaGaian wheel Spring Equinox is understood as a point of balance of the three qualities of Goddess, of Cosmogenesis, the

---

[431] Spretnak, *Lost Goddesses of Early Greece*, 105-118.

delicate balance of these qualities that keeps it all spinning. Persephone (as Seed) may represent both the Wise One from the depths (who is also Hecate) and the newly Emerged Young One, being embraced by the Mother. In this Moment there is rejoicing and affirming of the harmony of All. It is the three aspects of Cosmogenesis in "a fecund balance of tensions."[432]

This Seasonal Moment is often named as "Eostar" or some variation of it, from which the Christian rituals of Easter have taken their name since the Middle Ages of the Common Era; and the date for "Easter" – set for Northern Hemispheric seasons – is still based on the lunar/menstrual calendar. The name comes from the Saxon Goddess Eostre/Ostara.[433]

Some More Spring Equinox Story

The *full* story of Spring Equinox may be represented by a daffodil with bulb and roots exposed: that is, it includes the roots and grounding in the dark … the joy of the blossoming is rooted in the journey through the dark. Both Equinoxes celebrate this sacred balance, and they are both celebrations of the mystery of the Seed. The Seed is essentially the deep Creativity within, that manifests in the Spring as flower, or green emerged One. The flower is in the seed, the seed is in the flower. The lotus is present in the mud, the mud is present in the lotus – the lotus and the mud are present in each other: so it is for each one's personal journey, so it is for the collective journey, so it is for the human species, for the story of the Cosmos. The beauty that we enjoy and celebrate unfolds from the underworld.

Spring Equinox may celebrate the magic and joy of the unexpected, yet long awaited, green emergence from under the ground. It is understood as a Moment of certain emergence, and for enjoying the power of being: it is a "victory" if you like, and a good time for some contemplation on "power" – what it really is, its *organic* nature.[434] It is a

---

[432] Swimme and Berry, *The Universe Story*, 54.

[433] Walker, *The Woman's Encyclopedia of Myths and Secrets*, 267.

[434] As Judy Grahn describes "power" in "From Sacred Blood to the Curse and Beyond," in *The Politics of Women's Spiritualty*, edited by

power within every being, that all beings must have: it is Gaian power, the power of the Cosmos – unique and particular, and cosmic at the same time. Spring Equinox is marked on the PaGaian Wheel as a "Heraic Return," a term taken from Charlene Spretnak's re-storying of a hera as a courageous one.[435] The returning Beloved One, the Persephone, the Hera, is a redeemer figure: the Seed of Life that never fades away. It is a "resurrection" – the original: a celebration of the redeemer quality innate to Creativity, to Earth, and one in which each being participates, co-creates with their small, personal journeys and "victories"/returns.[436] This is Earth-based redemption, which we witness around us, are immersed in, and is expressed in the magic of Earth-Sun relationship.

The pre-"Olympic" games of Greece were Hera's games, held at Her Heraion/temple.[437] The winners were "heras" – gaining the status of being like Her. In the earliest of times Hera was not the bickering, jealous wife of the Olympian pantheon – She was Sovereign, the native/indigenous Queen. Hera, in the most ancient of stories was commonly identified with Gaia.[438] Marija Gimbutas tells that Hera was called the 'origin of all things,' by a Greek writer of the sixth century B.C.E.; that "Her name is cognate with Hora, season,"[439] and that Hera created and restored life. In recent research Marguerite Rigoglioso develops the theme of Hera's parthenogenetic capabilities and points to strong indications of a priestesshood *(parthenoi)* devoted to divine birth

---

Charlene Spretnak, 265-279. New York: Doubleday (1982), 265: "only an apple tree can make an apple, no one else can."

[435] Charlene Spretnak, "Mythic Heras as Models of Strength and Wisdom," in *The Politics of Women's Spirituality*, edited by Charlene Spretnak, 87-90: 87.

[436] See "The Equinoxes as Story of Redemption: Sacred Balance of Maternal Creativity" by Glenys Livingstone: https://pagaian.org/2016/09/20/the-equinoxes-as-story-of-redemption-sacred-balance-of-maternal-creativity-2/.

[437] Spretnak, *Lost Goddesses of Early Greece*, 87-88 referring to Harrison, *Myths of Greece and Rome*, (London: Ernest Benn Ltd., 1927), 18.

[438] Pamela Berger, *The Goddess Obscured* (Boston: Beacon Press, 1985), 16.

[439] Gimbutas, *The Language of the Goddess*, 134.

at Hera's sanctuary at Olympia: that the famed Olympic torch is "the burning flame of Hera *Parthenos* and her holy *parthenoi* that is carried around the world and guides the games until their safe conclusion."[440]

This celebration of Spring Equinox, as it has been done in PaGaian ceremony, is a "stepping into power," the power of being, as the light grows stronger and now dominant. All participants may identify themselves as *Courageous One*, perhaps entitling themselves as Hera, as Persephone, though other titles may be used, such as Hero or Shero.[441] Participants may express rejoicing in how they have made it through so much, having faced their fears, the chthonic, and their demise in its various forms. It is a time to welcome back that which was lost, and step forward into the light, with new wisdom, into the joy and power of being. Eostar/Spring Equinox is the time for enjoying the fruits of the descent, of the journey taken into the darkness. The young tender selves celebrated and committed to at Imbolc/Early Spring, are received into communal relationship and empowerment.

This may be understood as an individual experience, but also as a collective experience – as humans may emerge into a new Era as a species. Thomas Berry and Brian Swimme speak of the ending of the sixty-five million year geological Era – the Cenozoic Era – in our times, and our possible emergence into an Ecozoic Era.[442] They describe the Ecozoic Era as a time when "the curvature of the universe, the curvature of the earth, and the curvature of the human are once more in their proper relation."[443] Joanna Macy speaks of the "Great Turning" of our times.[444] Collectively most humans have been away from the Mother for some time. In the ceremony, each takes the torch/lantern and the seed, and wanders out of the circle (into the "Underworld") before coming to the "gate of emergence"; each may contemplate not only personal individual lost wanderings, but also that of the human species. Each

---

440 Marguerite Rigoglioso, *The Cult of Divine Birth in Ancient Greece* (New York: Palgrave Macmillan, 2009),138.

441 Max Dashu's term.

442 Swimme and Berry, *The Universe Story,* 241-261.

443 Ibid., 261.

444 See Macy and Brown, *Coming Back to Life.*

participates in a much bigger Return that is happening. Each may present themselves as the Beloved One returning to the Mother.

And as each one goes out of the circle and into the "Underworld," the rest of the ceremonial participants watch and wait – like Demeter. Each is Persephone, and each is Demeter. All welcome the lost Ones, the wiser returned Ones: each is the Mother and the Daughter-Self.

Swimme and Berry say that: "With the beginning of each new era of the universe, activity and its multiform possibilities undergo a creative transformation."[445] This is true of ourselves as well. Something completely new is possible – new synergies can come into being with the stepping onto a new platform, into a new strength of being. Swimme speaks of this as "space-time binding," where all the transformations of the past are held within the present moment, making a completely new leap possible[446] – into "what," we can only guess at. But the creative track record is good.

Spring Equinox/Eostar Breath Meditation

To begin this Seasonal Moment's contemplation:

> Centering for a moment … feeling the full cycle of your breath. With the inbreath, the light phase of the cycle, we may celebrate the Virgin/Young One aspect, the Urge to Be. With the outbreath, the dark phase of the cycle, we may celebrate the Old One aspect, She who creates the Space to Be. At the Solstices – with the peaking and full emptiness at the bottom of the breath, we may celebrate the Mother aspect, She who is this Dynamic Place of Being, the Sacred Interchange. At the Equinoxes, we may celebrate the Sacred Balance of all three – in relationship, that keeps it all spinning. Feel the full cycle of your breath – the delicate balance of all Three.
>
> Recognising the Triple Goddess Dynamic – the Creatrix – here in this primary Place … in your breath.

---

[445] Swimme and Berry, *The Universe Story*, 63.

[446] Swimme, *Powers of the Universe*, Program 9. "Interrelatedness."

Some Spring Equinox motifs

[Figure 20]

Suggested decorations for this Seasonal Moment, placed on the front door, windowsills, corners of shelves, and/or altars, may include:

- **green** – representing life on Earth, relationship with Sun, the advent of the chlorophyll molecule.
- **mask** – representing the journey and the emergence – the grief and the joy; or it may also be to represent identity with Persephone and/or Demeter.
- **flowers** – for the beauty and joy of being manifesting.
- **yellow gold ribbon** – associated with winning, perhaps with the Olympics and Hera's earlier games: especially appropriate if Australian (which has green and gold team colours).
- **coloured egg** – for new life, renewing power. Affirms traditional connection to "Easter" and Spring rituals – and in contrast to the seed of Autumn Equinox/Mabon.
- **a stone or gem** – representing the new Wisdom gained from the journey to the depths. My particular stones at Spring Equinox/Eostar point on the Wheel have a story of personal emergence … yours may too.

An Invocation

For months Persephone received and renewed the dead. She painted the forehead of each one and slowly pronounced:

You have waxed into the fullness of life
And waned into darkness;
May you be renewed in tranquility and wisdom.

Her Mother Demeter remained disconsolate and sat waiting for months. One morning a ring of purple crocus quietly pushed their way through the soil and surrounded Demeter. She looked with surprise at the new arrivals from below. She was too weakened to feel rage at Her injunction being broken. Then She leaned forward and heard them whisper in the warm breeze: "Persephone returns! Persephone returns!"

Everywhere the energy was stirring, pushing forth. Everywhere there was rejoicing – the Beloved One had returned, with new Wisdom from the depths.[447]

"Blessed are you – you have seen these things. You know the end of life and you know the divine –given beginnings."[448]

The Universe Story and Spring Equinox

Equinox is a celebration of the sacred balance – the creative tension in which life is born, the delicate balance in which Creativity of the Universe is possible. This creative tension may be celebrated at both Equinoxes. I understand the Equinox as a point of balance of the three qualities of Cosmogenesis in a fertile balance of tensions as described before.

Brian Swimme and Thomas Berry express the creative interaction:

> A cloud of elements hovered, floated ... far from the centre of the Milky Way galaxy. ...In our universe, the originating powers

---

[447] This invocational story is primarily the words of Charlene Spretnak, *Lost Goddesses of Early Greece*, 116-117, with my own summary.
[448] This quote is based on the words of Greek poet Pindar.

> permeating every drop of existence drew forth ten thousand stars from this quiescent cloud. To varying degrees, these stellar beings manifested the universe's urge toward differentiation, autopoiesis, and communion. And at least one of these, the Sun, managed to enter the deeper reaches of the universe creativity, a realm where the complexity, self-manifestation, and reciprocity at the very heart of the universe revealed themselves in a way transcending anything that had occurred for ten billion years – as an extravagant, magical, and living Earth burst into a new epoch of the universe story.[449]

That is, Earth Herself is a Spring Equinox/Eostar event. This epoch of teeming life as we know it is such a Moment in the greater scheme of things; expressed here by Swimme and Berry as the outcome of the interaction of the three qualities of Cosmogenesis – differentiation, autopoiesis, and communion and/or complexity, self-manifestation, and reciprocity.

Also as mentioned above, Persephone's emergence with new wisdom, with new power to Be, may be understood as a collective experience of emergence into a new era – the Ecozoic Era that Berry and Swimme speak of, and that Joanna Macy calls "The Great Turning." We may consider ourselves to be part of a much bigger ReTurn[450] that is happening – perhaps "returning to the Mother of us all."[451]

The Advent of the Chlorophyll Molecule

Brian Swimme describes something of the significance of the evolution of the chlorophyll molecule in "Canticle to the Cosmos;"[452] how essential it is for life on Earth. He also develops its connection with the retinol molecule in the eye.[453] This is about Earth's/our relationship

---

[449] Swimme and Berry, *The Universe Story,* 78-79.

[450] This spelling is inspired by the music of Jennifer Berezan's *ReTurning* CD, which has traditionally been used for PaGaian Spring Equinox ceremony.

[451] Jennifer Berezan, *ReTurning,* CD (2013). http://www.edgeofwonder.com/music/returning.

[452] *Canticle to the Cosmos,* video 2, "The Primeval Fireball."

[453] There is a relevant excerpt of this video on YouTube: "Sun's Energy

with Sun that creates the magic of Green and sight. Spring Equinox is the Seasonal Moment for particular celebration and contemplation of this unfolding, the *Greening* of Earth, and also the power of light entering the bodymind.

Notes on imagery and metaphors in the offered Ceremonial Script

The Call to Gather should be with strong dramatic drumming to herald the emergence and return from the Underworld, the "Stepping into Power of Being" as I have expressed the essential quality of this Seasonal Moment.

The focus of the Breath Meditation/Centering is on the experience of balance of light and dark, as Earth is so balanced in this Moment; imagining the breathing in as a swelling with light, and the letting go of the breath as a letting go into dark. It is spoken of as "a fertile balance of tensions" that at the Spring Equinox is visualised as about to tip into increased light, and at the Autumn Equinox is visualised as about to tip into deeper dark.

In Calling the Directions/Elements – Casting the Circle, the focus in this Season is on the power of being, the heraic return from the depths, thus on the elements as Powers, a naming of the elements as our true powers of life, as *Gaian Powers*, and our knowing of them – our certain sensing of them as powers that manifest in various ways. It continues and deepens the Early Spring/Imbolc focus on feeling/sensing the elements within the self, but now with an ability to act with them, an empowerment. The associations with each of the elements at this Seasonal Moment are based on psychologist Sarah A. Conn's four aspects of global responsibility – direct experience, understanding, action and awareness.[454]

For the Invocation, each participant holds up a special stone or rock representing the precious wisdom gained from the dark journey/ journeys: we are invoking the Hera/Hero/Shero, the Courageous One that each is. The placing of their stone on the altar is then an affirmation

---

and Relationship with Earth's Life,"
https://thegirlgod.com/pagaianresources.php.

[454] Sarah A. Conn, "The Self-World Connection," *Woman of Power* (Issue 20, Spring 1991), 73-74.

of their return and presence and gained wisdom.

The chosen food for Communion is poppyseed cake. The poppy is associated with Persephone for many reasons, and amongst them is its representation of the abundance and fertility of the dark, that may nurture the manifestation of strength of being. The chosen drink for Communion is a green juice of some kind, as a celebration of the power and magic of the chlorophyll molecule: the *Greening*. As each is offered the cake and juice they are blessed with an admonition commonly used in the Christian communion rite, which some participants would be aware of: "Do this in remembrance of She who gives Life" – it is a conscious reclaiming, as Persephone in this Goddess tradition is a Redeemer/Wisdom figure.

Each participant is also offered a "Golden Egg" (a gourmet chocolate wrapped in golden foil). These are meant to recall "the myths of Hathor-Astarte who laid the Golden Egg of the Sun,"[455] and we are celebrating the power of Sun. The stories of the "Easter Bunny" came from the Moon-Hare sacred to the Goddess – the stories got linked and interwoven over the millennia. Many primordial Goddesses laid Golden Eggs: that is, what they laid was the Sun, abundant and given freely everyday. For Southern Hemisphere people in particular, who celebrate Spring Equinox in September, the chocolate eggs calls to their sensed experience, association of "Easter" festivities with the appropriate Season: that is, with Spring instead of Autumn … it's a little bit of soul retrieval. A tradition of giving eggs at Easter developed as it evolved out of Spring Equinox celebrations. The blessing given with the offering of the golden eggs is a reminder that Sun's gift is given freely every day.

The suggested placing of a mirror in the Underworld for the Stepping into Power/New Wisdom rite is a connection to the Creation story from the Faery tradition where all manifestation springs forth from Goddess falling in love with Her reflection in the curved mirror of black space.[456] These ancients apparently understood that the essence of creative power springs from self-love, known and seen only completely in the dark: such self-knowledge is a gift of the darkness. After all have emerged from the Underworld journey rite and are seated quietly with

---

455 Walker, *The Woman's Encyclopedia of Myths and Secrets*, 267.
456 Starhawk, *The Spiral Dance*, 41.

their flowers and laurels, the words of Pindar, the Greek philosopher and poet, are used in part, for the blessing: "Blessed are you – you have seen these things…"[457]

Note that the Demeter-Persephone story told in the ceremony may be dramatized in many wonderful ways.

[Figure 21] A Spring Equinox/Eostar Ceremonial Altar

A SPRING EQUINOX/EOSTAR CEREMONIAL SCRIPT

Note that the directions here are called in "counter-clockwise" direction which is sunwise in the Southern Hemisphere. The order may be changed to "clockwise"/sunwise for the Northern Hemisphere. Note

[457] Jean Houston uses these words as participants in her exercise "The Realm of the Ancestors" emerge from the "Underworld." See Houston, *The Hero and the Goddess* (New York: Aquarian Press, 1993), 197.

that this particular offered ceremonial script is not considered prescriptive, and that there are many creative possible variations.

Some requirements and some suggestions: Each one bring a stone/rock, a Spring mask, a bouquet, head garland. Potted flower from Autumn Equinox/Mabon on altar. Centre large clay pot of soil, overlaid with wreath of greenery and flowers, with centre candle. Green altar cloth. A "gateway" from the circle to the "Underworld" – marked with stones each side. Seed, 3 sheaves of wheat, poppies, bouquets, garden torch/lantern near gateway. Stones (that each brings) in basket in the Earth direction on altar. Poppyseed ring cake, sliced. Green drink – perhaps chlorophyll in water. White wine and light colored juice. Chocolate eggs wrapped in gold paper (I use Ferrero Rocher). Music ready: I have used *ReTurning* by Jennifer Berezan. Boiled coloured eggs for decorations. Mirror in the Underworld.

Each one takes their special stone/rock to pre-ritual meditation with them, and then places it in the basket on the altar upon gathering to the circle. Head garlands will be left with Demeter at gate to Underworld for the journey; Demeter puts it back on head upon return.

Call to Gather ... drumming with a strong beat, until all are gathered, or a little longer.

Centering – Breath Meditation
This is the time of Spring Equinox, Eostar, in our Hemisphere, the moment of balance of light and dark in the light part of the cycle. The light and the dark in the South and in the North of our planet, are of equal length at this time.

Feel the balance in this moment – Earth as She is poised in relationship with the Sun. See Her there in your mind's eye. Feel for your own balance of light and dark within – this fertile balance of tensions. Breathe into it. Breathe in the light, swell with it, let your breath go into the dark, stay with it. Breathe in the light, swell with it. Feel for your centre … perhaps shifting on your feet, from left to right, right to left, feel for your centre ... breathe into it.

In our part of Earth, the balance is about to tip into the light. Feel the shift within you, see in your mind's eye the energy ahead, the light expanding. Feel the warmth of it. Breathe it in.

Statement of Purpose
This is the time of Spring's return. Warmth and growth may be sensed in the land, the flowers can be smelt in the air. The young light that we celebrated at Imbolc, has grown strong and come to balance with the dark. Life bursts forth with new strength. The story of Old tells us that Persephone beloved Daughter, returns from Her journey to the Underworld. Demeter, the Mother, stretches out Her arms – to receive and rejoice. The beloved One, the Lost One, returns with new Wisdom from the depths.

We may step into a new harmony. Where we step, wild flowers may appear; where we dance, despair may turn to hope, sorrow to joy, want to abundance. May our hearts open with the Spring.

Calling the Directions – Casting the Circle:
Let us begin our celebration by remembering our true power of life – our Gaian Powers, let us enter the eternal Space and Time in which we are immersed and embraced – the De-meter who holds us, who waits for us, who receives us.

DRUMS
Celebrant: Hail the East and Powers of Water ... powers of sensitivity, emotion and response: We are wet with you, we taste you, we know you.
Co-celebrant takes the water around the circle (sunwise) for to each wet their hands and wipe on themselves.
Response: I am wet with you, I taste you, I know you.

DRUMS
Celebrant: Hail the North and Powers of Fire ... powers of shaping, passion and action: We are spark of you, we feel your warmth, we know you.
Co-celebrant lights the fire and takes it around the circle; each passes their hand over it.
Response: I am spark of you, I feel your warmth, I know you.

DRUMS
Celebrant: Hail the West and Powers of Earth ... powers of presence, memory and understanding: We hold the story of you in our bodies, we feel your weight, we know you.
Co-celebrant takes the basket of stones/gems around the circle for each to take their own.
Response: I hold the story of you in my body, I feel your weight, I know you.

DRUMS
Celebrant: Hail the South and Powers of Air ... powers of perception, awareness and inspiration: We breathe you in, we expand with you, we know you
Co-celebrant lights the smudge stick and takes around the circle to each.
Response: I breathe you in, I expand with you, I know you.

Celebrant: This is the truth and power of who we are (PACING THE CIRCLE) Water, Fire, Earth, and Air. She is alive in us and we in Her. We are beyond the bounds of time and space, we are at Centre – the Centre is here (GESTURE to centre), where light and dark, birth and death, joy and sorrow meet as one.
(Celebrant light centre candle)

Invocation
Celebrant: Let us invoke/recognize the Sacred Heraic self in each other, that you are, the Courageous One who has journeyed and who journeys – who is emerging and present. Announce yourself.
Each person, holding up their rock/stone ... "I am a Hera, (or Shero or Hero), a Courageous One, Beloved One, who is returning, with new wisdom." AND/OR "… here, present with you." OR …
(option to say more – in terms of particular personal experience)
Each places their rock/stone on the altar.
Response from group: Beloved One, (repeat however they named themselves, and as much else as is remembered), we welcome you.

ALL SIT DOWN:
Celebrant: Courageous Ones, Heras, Persephones … who have gathered

here in this time and place, let us sit and listen to the traditional story of this time – versions of which have been told for aeons, and understood ever more deeply.

Story[458]

"Persephone had gathered three poppies and three sheaves of wheat. Then Demeter had led Her to a long, deep chasm and produced a torch for her to carry. She had stood and watched Her Daughter go down further and further into the cleft of the Earth. ...

For months Persephone received and renewed the dead without ever resting or growing weary. All the while Her Mother remained disconsolate. ... In Her sorrow She withdrew Her power from the crops, the trees, the plants. She forbade any new growth to blanket the Earth. The mortals planted their seed, but the fields remained barren. Demeter was consumed with loneliness and finally settled on a bare hillside to gaze out at nothing from sunken eyes. For days and nights, weeks and months She sat waiting.

One morning a ring of purple crocus quietly pushed their way through the soil and surrounded Demeter. She looked with surprise at the new arrivals from below and thought what a shame it was that She was too weakened to feel rage at Her injunction being broken. Then She leaned forward and heard them whisper in the warm breeze: "Persephone returns! Persephone returns!"

Celebrant, holding up a potted flower or gesturing towards potted orchids/ spring flower: "Persephone returns! The Seed of life never fades away. She is always present. She is returning from beneath to full flower. She does return from beneath to full flower. Beauty is restored."

"We are the Persephones – returning from beneath to full flower. Let us put on our masks, we are Her."
Each put on their Spring mask.

Chant and Dance (with tambourines and seed pods is good)
"Persephone Returns! Persephone Returns! Rejoice, rejoice!"
Option to link hands and grapevine steps, and/or with raised arms,

---

[458] Spretnak, *Lost Goddesses of Early Greece*, 114-117, combined with my own summary.

moving to outside space, if not there already.

Raising Energy
Celebrant: In the tradition of our ancestors and the people of this land too, let us join in the awakening of Mother Earth and ourselves.
Group does foot stomping on the Earth in time to: "She who is alive, is alive in us, and we who are alive, are alive in Her."

When stomping is done
Celebrant: Breathe up Her energy through your feet – Her electro-magnetic energy. Feel this Gaian Power.
STILLNESS WITH THE MOTHER
Celebrant: Let us move back to our circle, with this stillness (or the chant).
When the circle is all back and assembled:

Calling Demeter
Co-celebrant: "Let us call Demeter, the Mother – who holds us, waits for us and receives us."

All chant, with drums, and a facilitator to orchestrate (as celebrant changes headpieces, and adds any other Demeter garments):
"Demeter, Demeter – Mother we call you, Mother we call you."
LET ENERGY BUILD
When "Demeter" is ready and standing waiting, the group finishes with "Demeter!"

Demeter may walk the circle ceremoniously.

Stepping into Power/New Wisdom/ New Strength of Being
MUSIC ON – LOW
Demeter (standing near "gateway"): "Now is the time for you to step into the Power of Being, to leave behind the binds of the past, to welcome back that which was lost, and to welcome New Wisdom. Step forward one at a time, to make your journey, enter the Underworld – to remember something of your Journey and your emergence, your Return. We will wait for you."

Co-celebrant goes to the gateway to light the lantern.

MUSIC – VOLUME UP
EACH LEAVE GARLAND WITH DEMETER, SHE PUTS IT BACK ON HEAD UPON RETURN
Each one in turn, goes out of the circle and takes the seed (and/or wheat and/or poppies), takes the lantern and wanders into the Underworld.

IN THE DARK UNDERWORLD, THERE IS A SMALL TABLE WITH A MIRROR, INTO WHICH THE JOURNEYER MAY LOOK – ENACTING PART OF THE PURPOSE OF SUCH JOURNEYING: that is, self-knowledge, the deep source of Creativity, the autopoiesis of the Cosmos within the self.

The group plays the part of "Demeters" watching and waiting for the wandering Persephone.
Upon return to the "gateway" of the circle the Journeyer puts lantern down and stands or sits there.

PAUSE MUSIC

Each: (Optional words offered … or each write their own)
"I am a Beloved One returned/returning. I have gained new Wisdom. I feel new knowings. I have returned from … I have new strength and power – Gaia's Power. I leave behind the binds of the past – they are for me ...
I welcome (back) my beloved Daughter/Child within … the Beloved Lost One. She/This One is for me ...
I step with Her/Him/Them into the light & strength of Spring & Being."

Each steps through the gateway, takes off their mask, is embraced by Demeter and given a bouquet. Demeter puts their head garland back on. Others may embrace the returned One too. Demeter may speak welcoming words, such as: "You are a champion, a hera, welcome …"
Group response: cheer and clap!! "What a hera! We bless your empowerment. You made it!" or something appropriate.

(Switch music back on, and await next person to take the torch/lantern)

When all have journeyed and returned
Sitting
Celebrant: Let us sit quietly for a while – rest on our laurels!
(optional music – a harp is good)

Celebrant: Blessed are you – you have seen these things. You know the end of life and you know the sacred beginnings. May you enjoy the wisdom of your journey.

Communion
Celebrant (holding up the cake): Take and eat this holy cake – with its many seeds, remembering Her fertility and abundance. Do this in remembrance of She who gives Life.

Celebrant pass the plate of cake sunwise. Each repeat blessing as they hand to next:
"Take and eat, do this in remembrance of She who gives Life."

Server take glasses to all.

Celebrant (holding up the green drink): Blessed are you – you have seen these things. Take and drink this Greenness, the magic of the chlorophyll molecule and Earth-Sun communion – do this in remembrance of She who gives Life.
Celebrant pass decanter of green drink sunwise. Each repeat the blessing as they pour for the next:
"Take and drink – do this in remembrance of She who gives Life."
Offer white wine or juice, and more cake.

Stories (with ceremonial manners)[459]
Celebrant: Let us tell each other stories of empowerment, and anything else you would like to share.

---

[459] These manners require listening with no comment or additions to the speaker's story, and confidentiality.

Possible response to each storyteller/speaker: "We bless your empowerment."

Offer Eostar eggs, around the circle:
"Take the Golden Egg that the Ancient One lays out for you every day – it is yours, freely given."

Open the Circle
Celebrant invites all to open the circle, as all turn to the South:
We have remembered that we are Air, we are the breeze, She expands in us. May there be peace within us.
All: May there be peace within us. (each may gesture with folded arms over the heart or as they wish)

Celebrant (all turning to the West): We have remembered that we are Earth, we are the story, She speaks in us. May there be peace within us.
All: May there be peace within us.

Celebrant (all turning to the North): We have remembered that we are Fire, we are the dance, She is our form. May there be peace within us.
All: May there be peace within us.

Celebrant (all turning to the East): We have remembered that we are Water, we are the juice, She feels with us. May there be peace within us.
All: May there be peace within us.

Celebrant (all turning to the centre): We have remembered that we are Sacred Ones – Heras, Persephones, Courageous Ones who have Journeyed and ReTurned, and are ReTurning. We have remembered something of our Journeys, our Lost Places. We have welcomed Lost Beloved Ones. We have welcomed and embraced New Strength of Being, New Wisdom. May there be peace within us and between us.

All: May there be peace within us and between us (all join hands).

Celebrant: The circle is open but unbroken, may the Peace, Wisdom and Courage of Goddess go in our hearts and minds.

All: It has been a merry meeting, it is a merry parting. May we merry meet again. Blessed Be!

Music: suggested "Shine" by Wendy Rule.[460]

Points for individual contemplation prior to the ceremony:

- the balance of this Moment, the light and dark in balance on the planet and within you … this fertile balance of tensions.
- yourself as a Hera/Shero/Hero/Courageous One … some of what you have made it through, how you have returned from the depths with Wisdom – represented with the rock/stone (take it with you for contemplation).
- stepping into Power of Being – what that is for you, your journey, what/whom you are welcoming, perhaps what 'binds' of the past you are leaving behind, how you might like to express this (spontaneous or written).

Spring Equinox/Eostar Goddess Slideshow:
https://thegirlgod.com/pagaianresources.php

---

[460] Wendy Rule, *Deity* CD (1998).

CHAPTER 8
BELTAINE/HIGH SPRING

Southern Hemisphere – October 31st or 1st November
Northern Hemisphere – April 30th (May Eve) or 1st May
These dates are traditional, though the actual astronomical date varies. It is the meridian point or cross-quarter day between Spring Equinox and Summer Solstice, thus actually a little later in early November for S.H., and early May for N.H., respectively.

Beltaine/High Spring may be understood and felt as a celebration of the *high point* of the light phase of the yearly cycle (though it is not actually), because the light part of the day is continuing to increase at this point and will not yet turn as it will at Summer Solstice; that is, light is full-on at this Seasonal Moment, without any turning in to dark ... light is uncompromised.[461] In this cosmology Beltaine/High Spring is still associated with *She Who is the Urge to Be*, the Young One/Virgin aspect of the Triple Goddess; but at this late developed stage of the light cycle, as She steps into power of being (Spring Equinox phase), passion and desire for life leads Her to engagement with Other, the fruiting begins. Whereas at Imbolc/Early Spring, which is the first celebration of this Young/Urge to Be quality, in the light phase of the year's cycle, when there is an *identification* with this quality in each differentiated form,[462] at this Moment of Beltaine it is a clear participation in Her *process* of the Dance of Life, the celebration of Her passion, as exuberant life unfolds.[463] The Young

---

[461] Just as Samhain/Deep Autumn may be felt and understood as the deep point of dark: that is, dark is full-on, it is not turning into light as it will be at Winter Solstice. See Chapter 2 "In Summary: Contemplating How Creativity Proceeds."
[462] See Chapter 6 of this book.
[463] Just as Lammas/Late Summer, the first celebration of the dark phase of the cycle may be an *identification* with *She Who Creates the Space to Be* quality – how you are this, and Samhain/Deep Autumn may be a participation on Her *process* of transformation, the celebration of Her

One has matured and comes into relationship with other, and is morphing into the Mother quality: it is the *communion* quality of Cosmogenesis that is dialling in. Whereas at Imbolc/Early Spring the Young One was coming forth *from* the Mother, now at Beltaine/High Spring She is changing *to* the Mother/Creator, *becoming* this fullness of being.

There is a lot of "exoteric hoo-haa" about Beltaine, which you may encounter and even enjoy; and by that I mean popular and quite superficial interpretations of the significance of the Moment. But like all the "native seasonal festivals" of this Celtic/pre-Celtic calendar, it may "be approached on a much deeper level."[464] Traditionally in Pagan circles in relatively recent periods of human history the passion and desire for life has been celebrated almost exclusively in terms of female and male biology, frequently expressed as "the Goddess and the God," especially so in the modern era when this emphasis and pre-occupation may be more the result of modern Freudian and Jungian influence. In the interest of essential Cosmic Creativity, meiotic sex (hetero-sex) does indeed seem to have been a positive move: it exponentially increased diversity and complexity in the Cosmic unfolding, and hence also communion. Yet allurement and passion are not bound to this particular engagement, nor are they apparently dependent upon it: that is, Creativity of the Cosmos proceeded long before meiotic sex. Mystical traditions have used the metaphor of "Beloved and Lover," which opens up many valencies to the experience of desire, passion and allurement.[465] Cosmologist Brian Swimme says that the ultimate urge represented by sexuality is "identification with the pulse of Creativity itself."[466] which is worthy of much more contemplation. The "pulse of Creativity" may be interpreted

---

fertile compost for renewal. See Chapter 2 "In Summary: Contemplating How Creativity Proceeds."

464 Matthews, *The Western Way*, 47.

465 See Livingstone, *PaGaian Cosmology* 60 or 57-62, for more on the terms 'feminine' and 'masculine'. Also see Starhawk's 1999 edition of *The Spiral Dance*, where she qualifies a lot of traditional Pagan expression for polarities: 267-268 on "female-male polarization" in particular.

466 *The Earth's Imagination*, program 6, "An Ocean of Energy."

more deeply and is older and deeper than meiotic sex.

Traditionally in PaGaian Cosmology, I have associated Beltaine with Aphrodite in particular, but it could be Venus, Ishtar and/or other deities from other cultures who are understood to embody the primordial, essential and ecstatic power of allurement/attraction, of erotic power:[467] a power that holds all things in form and unites the Cosmos,[468] the "glue" that allows and brings forth the Dance of Life. Nothing happens without this primary power, and I write it as Allurement, inspired by Brian Swimme's naming of it as a power of the Universe.[469] In a presentation called *Cosmology of the Goddess* given by Charlene Spretnak and Brian Swimme in 1987 at an Ecofeminist Conference, Charlene Spretnak explained the pre-Olympian history of Goddess myths and the original nature of Aphrodite,[470] and Brian Swimme spoke of the scientific principle (Mach's Principle) that recognises gravitation as the mysterious attraction binding the universe.[471] These insights have deeply influenced PaGaian poetry for this Seasonal Moment of Beltaine.

Some More Beltaine/High Spring Story

Earth's holy day of Beltaine/High Spring marks the meridian point of the lightest quarter in the cycle. Even though light has not yet peaked, the hours of light are longer than the hours of dark and are continuing to lengthen. Beltaine is polar opposite Samhain/Deep Autumn on the Wheel of the Year, when the dark is still climaxing; and with practice and consciousness, the relationship of the two Seasonal Moments can be noticed. Indeed it is said traditionally in Pagan circles that at both Moments the "veil is thin that divides the worlds," an

---

467 See Lorde, "Uses of the Erotic: The Erotic as Power," in *Weaving the Visions*, 208-213.
468 As Aphrodite is praised in the Orphic Hymns: http://www.asphodel-long.com/html/orphic_hymns.html.
469 Swimme, *The Powers of the Universe*, Program 2, and also *The Universe is a Green Dragon*, 43-52.
470 See Spretnak, *Lost Goddesses of Early Greece*, 69-72 for her telling of the pre-Olympian story of Aphrodite.
471 Spretnak, *Lost Goddesses of Early Greece*, xv –xvii.

expression of an *intimacy,* as I understand it. Beltaine's occupation is with the fertility and manifestation/creating of form – manifest reality, frequently a genetic fertility. Samhain's occupation is with the fertility of the imaginal depths and the decomposition of form – the manifesting reality.

At the time of Beltaine, in that hemisphere of the world, Earth continues to tilt further toward the Sun, source of Her life and pleasure. It has been felt and understood in Indigenous Pagan traditions as the entry into Summer: in pastoral economies in colder climates Beltaine was the time when cattle were let out of the barns (and correspondingly Samhain is understood as the official *end* of Summer: the name *Samhain* means "Summer's end"). The tender Promise of being of Early Spring has grown strong at this time of Beltaine/High Spring, is empowered with wisdom from the depths (celebrated at Spring Equinox), and is drawn with deep desire into the dance of life, waxing towards fullness of being.

The name "Beltaine" is commonly said to mean 'bright fire,'[472] and it is common to associate the name with a rumoured Celtic god Beil or Bel. But the evolution of the term is probably much older and more complex. There is ancient association of 'fire' with 'dea'/goddess, of elemental fire and divine eye, and *Aine* is the name of a Celtic Sun Goddess. In the old Irish language, 'eye' and 'sun' are the same word – *suil.*[473] There is common Indigenous recognition of the power of the Sun entering the eye: it is said "the Sun is in the eye."[474] And the retinol molecule of the vertebrate eye is formed by the same processes as the chlorophyll molecule, which synthesises light.[475]

According to Goddess scholar Michael Dames, *Belt* means 'white,' perhaps with the same root as the Lithuanian word 'baltas' meaning 'white,' and thus may refer to the white lunar eye.[476] So together with *Aine* the name of the primary Sun Goddess, "The word ... Beltaine,

---

472 For example: see Crowley, *Celtic Wisdom*, 80.
473 Dames, *Ireland,* 196.
474 I learned this from Brian Swimme in his *Canticle to the Cosmos* series.
475 Swimme speaks of this relationship in *Canticle to the Cosmos*, video 2, "The Primeval Fireball."
476 Dames, *Ireland,* 199.

may bring the eyes of night (Moon) and day (Sun) together:"[477] that is, the eye of the dark (and Winter) and the eye of the light (and Summer). The traditional red and the white colours of this season then may have as much to do with relationship of Moon and Sun – *the perceived power of Moon and Sun,* Winter to Summer … and one could say of Beloved and Lover – not necessarily "sexual" (or heterosexual) in the usual meaning of the word, but certainly erotic, ecstatic, merging. In Ireland the Seasonal Moment is often named as Bealtaine, which is also part of the Irish word for the month of May.

The twin fires lit in older times on hilltops in Ireland for Beltaine likely represented the two eyes of night and day.[478] With this vision, Goddess as Sun and Moon sees Her Land, and with the power of Her eyes (Sun and Moon) brings forth life and beauty. With the fire eyes, Goddess "reoccupied and *saw* her whole land…"[479] The twin fires later came to be used to run cattle between as they headed out to Summer pasture, for the purpose of burning off the bugs and ticks of Winter; the fires may thus be understood to serve a cleansing effect and likely the origins of the tradition of the ceremonial leaping of flames by participants in Beltaine festivities. In PaGaian Cosmology this is poetically expressed as the Flame of Love that burns away the psyche's "bugs and ticks," and *sees* the Beauty present, and calls it forth. The Beltaine flames may be a celebration of Sun entering into the eye, into the whole bodymind: a powerful creative evocation upon which the Dance of Life depends, and as the cleansing power of love and pleasure.

PaGaian focus for Beltaine is on the Holy Desire/Passion for life, and it may be accounted for on as many levels as possible … the complete holarchy/dimensions of the erotic power. On an elemental level, there is our desire for Air, Water, the warmth of Fire, and to be of use/service to Earth. There is an essential longing, sometimes nameless, sometimes constellated, experienced physically, that may be recognized

[477] Ibid., brackets mine. This is one analysis of the word Beltaine … there are other conjectures: none are certain it seems, so Poetry of the place and time seems acceptable.
[478] Ibid., 195-199.
[479] Ibid., 196.

as the Desire of the Universe Herself – desiring in us.[480] We may remember that we are united in this desire with each other, with all who have gone before us, and with all who come after us – all who dance the Dance of Life. Beltaine is a time for dancing and weaving into our lives, our heart's desires; traditionally the dance is done with participants holding ribbons attached to a pole or tree (a Maypole in the Northern Hemisphere, which may be renamed as a "Novapole" in the Southern Hemisphere), wrapping the pole with the ribbons. This is not simply the heterosexual metaphor as is thought in modern times (thanks largely to Freudian thinking) – it is deeper than that. As Caitlin and John Matthews point out: it is

> symbolic of a far greater exchange than that between men and women – in fact between the elements themselves. ... the maypole, a comparatively recent manifestation in the history of mystery celebrations, can be seen as the linking of heaven and earth, binding those who dance around it ... into a pattern of birth, life and death which lay at the heart of the maze of earth mysteries.[481]

Beltaine is a celebration of Desire on all levels – microcosm and on the macrocosm, the exoteric and the esoteric.[482] It brought you forth physically, and it brings forth all that you produce in your life, and it keeps the Cosmos spinning. It is felt in you as Desire, it urges you on. It is the deep awesome dynamic that pervades the Cosmos and brings forth all things – babies, meals, gardens, careers, books and solar systems. We have often been taught, certainly by religious traditions, to pay it as little attention as possible; whereas it should be the cause of much more meditation/attention, tracing it to its deepest place in us. What are our

---

480 I have been inspired and informed by Swimme's articulations about desire, particularly in *Canticle to the Cosmos,* video 2 "The Primeval Fireball," video 5 "Destruction and Loss," and video 10 "The Timing of Creativity."

481 Matthews, *The Western Way*, 54. And for more, see "Creativity of Beltaine Moment": http://pagaian.org/2014/10/30/1279/.

482 Desire is a big topic: see Livingstone, *PaGaian Cosmology,* 251-253 for more.

deepest desires beneath our surface desires. What if we enter more deeply into this feeling, this power? It may be a place where the Universe is a deep reciprocity – a receiving and giving that is One. Brian Swimme says, in a whole chapter on "Allurement":

> You can examine your own self and your own life with this question: Do I desire to have this pleasure? Or rather, do I desire to become pleasure? The demand to 'have,' to possess, always reveals an element of immaturity. To keep, to hold, to control, to own; all of this is fundamentally a delusion, for our own truest desire is to be and to live. We have ripened and matured when we realize that our own deepest desire in erotic attractions is to become pleasure ... to enter ecstatically into pleasure so that giving and receiving pleasure become one simple activity. Our most mature hope is to become pleasure's source and pleasure's home simultaneously. So it is with the allurements of life: we become beauty to ignite the beauty of others.[483]

Beltaine is a good time to contemplate this animal bodymind that you are: how it seeks *real* pleasure. What is your real pleasure? Be gracious with this bodymind and in awe of this form, this wonder.

Beltaine is also a good time to contemplate light, and its affects on our bodyminds as it enters into us; how our animal bodyminds respond directly to the Sun's light, which apparently may awaken physical desires. Light vibrates into us – different wavelengths as different colours – and shifts to pulse. It is felt most fully in Springtime ("spring fever"), as light courses down a direct neural line from retina to pineal gland. When the pineal gland receives the light pulse it releases "a cascade of hormones, drenching the body in hunger, thirst, or great desire."[484] We respond directly to Sun as an organism: it is primal.

Beltaine/High Spring Breath Meditation

To begin this Seasonal Moment's contemplation:

---

483 Swimme, *The Universe is a Green Dragon*, 79.

484 Laura Sewall, "Earth Eros, Sky," *EarthLight* (Winter 2000), 22.

Centering for a moment … feeling the full cycle of your breath. Recognising the Triple Goddess Dynamic here in this primary Place: She who unfolds the Cosmos … receiving the Gift, as you inhale – feeling the Urge to Be, feeling the peaking – the exchange, the communion that this Place is, then the release – the dissolving, becoming the Gift, making space, ever transforming: the Triple Spiral dynamic here in your breath, the Wheel of the Year in your breath.

Some Beltaine/High Spring motifs

[Figure 22]

Suggested decorations for this Seasonal Moment – on the front door, windowsills, corners of shelves, and/or altars, may include:

- **the red** – commonly understood in more recent times to represent the fertility of the maturing female: the menarche, the first menstrual flow. An earlier significance, and another layer to it, is that it represents the fire of Sun – and in Ireland where the name Beltaine has its origins (written as "Bealtaine" traditionally), the Sun Goddess was known as *Aine* as already noted, but it may represent any Sun Goddess from around the globe.
-

- **the white** – commonly understood in more recent times to represent the fertility of the maturing male: his first flow. Another layer and perhaps an earlier understanding is that the white represents the light of Moon, as noted earlier. Beltaine is understood as the transition Moment from Winter to Summer: thus the light of Moon to light of Sun, night to day: Her twin eyes come together then in deep creativity. The weaving of the red and the white at Beltaine may represent participation in the 'pulse of Creativity' itself – extant in the creativity of Sun and Moon in relationship, with Earth.
- **the Maypole/Novapole** – may represent (or be) the sacred tree, the growth of new vegetation. It may be understood as representing cosmic/earth regenerative energy, with various valencies ... tree, spine, earth-cosmos connection. In *The Language of the Goddess*, Marija Gimbutas refers to column-like art and phallic motifs as "columns of life" and "cosmic pillars."[485] The dance around it involves both sunwise and counter-sunwise movement – creation and dissolution.
- **the rainbow streamers** – may represent how light breaks up into the full spectrum of colours, and the diverse creativity of life; as well as specific colours perhaps signifying elements, places, emotions and more – qualities that participants may desire to weave into their lives.
- **Honeypot** – sweetness of life, which is tasted at this time when sweet desire for life is met, and the fruiting begins.
- **Seashell** – represents Aphrodite, whom I understand to be one of the names and forms given to the primordial essence of Holy Desire by the ancients, as noted earlier.
- **the Flame(pot)** – may represent the "fire-eye" of the Goddess. The Beltaine flames may be a celebration of Sun entering into the eye, into the whole bodymind; this powerful creative evocation upon which the Dance of Life depends, and which may be seamlessly connected to the cleansing power of Love and pleasure.

---

[485] Gimbutas, *The Language of the Goddess*, 221-235.

- **Rose/flower or an object of beauty for you:** represents (is) the beauty called forth, and present in each one. In the Beltaine ceremony we may practice seeing beauty in each other and ourselves – invoke it: where there is perception of beauty there is love, and allurement into relationship, communion, into Creativity.

An Invocation

Love begins as allurement – as attraction. Think of the entire cosmos: ... Gravity is the word used by scientists and the rest of us in the modern era to point to this primary attraction. ... The attracting activity is a stupendous and mysterious fact of existence. Primal. ...[486]

Aphrodite … who is allurement itself,
"Gravity" as You have been named.
Primordial Essence!
You hold all things …
gather all things, sweep all, draw all …
into Yourself
with your Desire.

May we live more erotically …
true to our nature.
All hail Holy Lust of the Cosmos!
May we be renewed with you,
and experience your Joy and Ecstasy.

The Universe Story and Beltaine

At Beltaine, the essential primordial power of Allurement is celebrated. It is this power of attraction, in particular as gravitational bonding, that holds the universe together, as Brian Swimme describes it.[487] "This primal dynamism awakens the communities of atoms, galaxies, stars, families, nations, persons, ecosystems, oceans, and stellar

[486] Swimme, *The Universe is a Green Dragon,* 43-45.
[487] Swimme, *The Powers of the Universe*, Program 2, "Allurement."

systems."[488] In the experience of the ceremony, individuals may sense and express their participation in this Allurement, in its many valencies – feel and affirm how this Desire is at the core of their Being, and all that they do.

If desire/allurement is the same cosmic dynamic as gravity as Swimme suggests,[489] then desire like gravity is the dynamic that links us to our place, to "that which is," as philosopher Linda Holler describes the effect of gravity.[490] Held in relationship by desire/allurement we lose abstraction and artificial boundaries, and "become embodied and grow heavy with the weight of the earth."[491] We then know that "being is being-in relation-to"; Holler says that when we think with the weight of Earth, space becomes "thick" as this "relational presence … turns notes into melodies, words into phrases with meaning, and space into vital forms with color and content, (and) also holds the knower in the world."[492] Thus, ***I*** at last become a particular, a subject, a felt being in the world – a Place laden with content, sentient.

One of the shaping powers of life is a wild energy. Swimme and Berry speak of three biological shaping powers that correspond to the three qualities/faces of Cosmogenesis: genetic mutation, natural selection and conscious choice.[493] I associate the wild Virgin/Young One energy of Beltaine with the power of genetic mutation; Swimme and Berry say: "Genetic mutation refers to spontaneous differentiations taking place at life's root."[494] Wild energy is also associated with Artemis in Her Virgin aspect;[495] this is not the Virgin of patriarchal story. She and many other Goddesses were named as "Lady of the Beasts." Swimme

---

488 Swimme, *The Universe is a Green Dragon*, 49.
489 Ibid., 43.
490 Linda Holler, "Thinking with the Weight of the Earth: Feminist Contributions to an Epistemology of Concreteness," *Hypatia,* Vol. 5 No. 1, Spring 1990: 2.
491 Ibid., 2.
492 Ibid.
493 See "The Three Biological Shaping Powers and the Female Metaphor/Dea" in Chapter 1.
494 Swimme and Berry, *The Universe Story,* 125.
495 Spretnak, *Lost Goddesses of Early Greece,* 75.

and Berry describe the wild power of mutation as "a face ultimacy wears," "a primal act within the life process."[496] They say:

> A wild animal, ... alert and free, moves with a beauty ... far beyond the lock-step process of a rationally derived conclusion. The wild is a great beauty that seethes with intelligence, that is ever surprising and refreshing ... The discovery of mutations is the discovery of an untamed and untameable energy at the organic centre of life. ... For without this wild energy, life's journey would have ended long ago.[497]

Beltaine is a traditional time for lovers to marry in this Earth tradition, since this season especially celebrates passion; it is usually called "handfasting." It is also a time when *new* partners may be taken – a *trans*personal celebration of sexuality. In terms of the story of the Universe, Beltaine may be a celebration of the *advent of meiotic sex,* one point five billion years ago (see Appendix C): that is, the advent of the male/gendered sex, as it may be understood. Until that moment, all being may be understood as female, since it was the female egg that divided parthenogenetically, in the process of mitosis. This advent was a significant moment of Gaia's story/Cosmogenesis, in the Creative Unfolding of the Universe; it took a deeper leap into differentiation, communion and subjectivity.[498] That is one dimension of the "Novapole" rite, when we dance the Dance of Life. It may in part be a celebration of the Creativity represented in the fertility of meiotic sex in its Cosmic significance, its advent and contribution to Cosmogenesis. This moment in the Story of the Universe is also connected to the advent of death,[499] which is poetically resonant with the polar symmetry of Beltaine and Samhain.

And a further quote from *The Universe is a Green Dragon,* that is resonant with Beltaine/High Spring:

---

496 Swimme and Berry, *The Universe Story,* 125.
497 Ibid., 127.
498 Ibid., 107-109
499 See Sahtouris, *Earthdance*, 134-135, and Goodenough, *The Sacred Depths of Nature*, 143-149.

> Think of the entire cosmos, all one hundred billion galaxies rushing through space: At this cosmic scale the basic dynamism of the universe is the attraction each galaxy has for every other galaxy. ... Gravity is the word used by scientists and the rest of us in the modern era to point to this primary attraction. ...(but) the mystery remains no matter how intelligently we theorize. ... We awake and discover that this alluring activity is *the* basic reality of the macrocosmic universe ... this alluring activity permeates the cosmos on all levels of being. ... By pursuing your allurements, you help bind the universe together. The unity of the world rests on the pursuit of passion.[500]

Notes on imagery and metaphors in the offered Ceremonial Script

For the Call to Gather when possible, a conch shell made into a horn, is great, as it may be representative of Aphrodite (or Her sister primordial oceanic deities) who may be especially invoked at this Seasonal Moment. Otherwise, vigorous drumming that may reflect the fertile alluring beat of life.

The focus for the Centering – the Breath Meditation begins with feeling the breath as it waxes towards fullness and feeling one's desire for this breath. There is recall that we are present to all who have gone before us in the one deep breath each takes: this is not just poetry, it is empirical fact apparently.[501] We breathe deeply and contemplate the presence of all those who have gone before, and our presence to all those who come after us: this is also acknowledgement of the intimacy of death in the Moment, a recognition of the resonance with the polar opposite Seasonal Moment of Samhain, as noted earlier. There is opportunity and suggestion to make a simple physical gesture to embrace this presence and the present moment.

For Calling the Directions – Casting the Circle: since this is a celebration of Desire/Allurement, the focus is on our experience of desire for each of the elements – our thirst, desire for warmth and gathering around, our desire for fulfilment of purpose, and for breath and inspiration. Each participant engages in a tactile sensuous way with

---

[500] Swimme, *The Universe is a Green Dragon,* 43-48.

[501] Swimme, *Canticle to the Cosmos*, video 1, "The Story of Our Time."

each element. There is an identification of that desire with all desire, the desire of all who have come before us, and the desire of all who will come after us, and with the Desire of the Universe. And we remember primarily the *presence* of the elements; that is, that our desire is met … we are not left wanting, it is consummated. We didn't make up this desire, it is the Universe desiring in us.[502] It is our nature. We are united in these desires; that is, we all experience these desires. And we are united in our desire with all others who have gone before and all who will come after, in their desires.

The ceremonial Invocation presents an opportunity for participants to identify with something that they find particularly beautiful – the beauty of the flower (the rose is Aphrodite's flower of choice, though a native flower may be desirable), tree, bird, ocean … it may be a place, something that reveals for each the beauty at the heart of the Universe. Each is encouraged to identify with this beauty, to identify themselves as "the Beauty whom She desires." This is an expression that Brian Swimme has used to describe what he perceives as the required aspiration of the male in sexual relationship to the female; that is, to be "the beauty that she desires."[503] I like to use it here meaning it as something which all Being *is* actually, and all may consciously aspire to; that is, to be the beauty and delicious morsel that the Universe desires. To facilitate this identification of self with beauty, each one may bring an object or a photo. This may be extended, as each one wishes, to expressing themselves as the Beloved, wholly desirable, and the holiness of their desiring being.[504]

The rite of Leaping the Fire is grounded in old traditions as mentioned earlier, and storied here in ceremony as "letting the *Flame of*

---

502 An expression from Swimme, *Canticle to the Cosmos*, video 5, "Destruction and Loss."

503 In *The Earth's Imagination*, Program 3, "Synergy."

504 At a subtle level this celebrated Desire is a desire for mergence – union. With practice of the Wheel of seasonal ceremony one begins to get a SENSE of growing light associated with this dissolution of union – we are actually moving towards entropy with the waxing and alluring light. This gets "written" into the bodymind with this ceremonial practice, though it may be unconscious scribing for some.

Love" burn away "petty disharmonies and habitual negativities,"[505] and leaving behind what we will.

For Communion it is the sweetness of life that is emphasized, with honey cake and sweet wine/juice being offered, with the admonition to "consummate your desire" and to enjoy it. In PaGaian tradition it is a pink ring cake (a "Yoni-cake"[506]) with rose water and honey on it that is dramatically presented and consumed. With the drinking of sweet wine/juice (pink preferably), each is encouraged to enjoy the sweetness of life, not to miss the Moment.

[Figure 23] A Beltaine/High Spring Ceremonial Altar

A BELTAINE/HIGH SPRING CEREMONIAL SCRIPT

Note that the directions here are called in "counter-clockwise" direction which is sunwise in the Southern Hemisphere. The order may be changed to "clockwise"; sunwise for the Northern Hemisphere. Note that this particular offered ceremonial script is not considered prescriptive, and

505 An expression of Charlene Spretnak's: I can't remember where.
506 See Appendix G for a recipe.

that there are many creative possible variations.

Some requirements and some suggestions: Ceremonial altar as it is scripted here is in a different location to the two firepits/cauldrons required and a tree hung with rainbow ribbons – it need not be. Pink ring cake, topped with rose water and honey and petals: the cake sliced but whole. Sweet pink wine/juice and champagne. Variety of berries and/or heart-shaped chocolates. Rose water in the East. Firepot[507] in the North. A large dinner plate of wet earth and card paper for handprints in the West, with bowl of water and towel. A small bouquet of scented flowers and/or herbs in the South. Participants may bring garlands. Each may bring an object of beauty – rose/flower/gemstone/ whatever, or a photo, for the invocation. Music ready for invocation.

Call to Gather … with drumming or a conch shell

Centering – Breath Meditation

Breathe deep. Feel your breath as it rushes in, as it waxes towards fullness. Feel your desire for this breath, for this fullness. Breathe deep and feel your breath as it waxes towards filling completely. Feel your desire for this breath. This is how Gaia breathes in our part of the world at this time. The light is waxing towards fullness. Feel your breath as it waxes towards fullness. Feel your desire for this breath, this fullness ... Earth's desire for this breath, for this fullness of being.

And again, taking a deep breath … in that breath we share in the life of all who have come before us, and danced the dance of life: & we are present to all who will come after us and dance the dance of life – breathe deep … and feel the Presence of all these … and the Present as it always is: the Gift that it is. You may if you like make a simple gesture to embrace this Presence and to receive this "Present" – reaching back and down – to all who have come before us, bringing your hands and arms up and forward – reaching to all who will come after us, and folding over

---

507 This may be a clay pot of sand into which a small amount of methylated spirits will be poured and lit. It produces a soft flame that will not set off fire alarms, though care should still be taken.

in reverence as you embrace it all. Breathing. And again … as you like – reaching back and down, harvesting the ancestors, bringing your arms up and forward, reaching into the future to the descendants of your body-mind-spirit, and folding your arms over, embracing, holding this Presence. Breathe deep – in this Present Moment, that You are. We are the Present Moment. Feel your desire for this.

Statement of Purpose

This is the Moment of Beltaine, when the light part of the day is longer and continues to grow longer than the dark part of the day. Light is waxing towards fullness. In our region, Earth continues to tilt us further toward the Sun – the Source of Her pleasure, life and ecstasy. This is the time when the sweet Desire for Life is met. The fruiting begins. It is the celebration of Allurement … Holy Lust ... that which holds all things in form and allows the dance of life.

The ancients called this Holy Lust, this primordial Essence "Ishtar," "Venus" – "Aphrodite" … they sang of Her:

> "For all things are from you,
> Who unites the Cosmos.
> You will the three-fold fates.
> You bring forth all things.
> Whatever is in the Heavens.
> And in the much fruitful earth
> And in the deep sea."[508]

Let us celebrate our erotic nature, which brings forth all things.

Calling the Directions – Casting the Circle:

We may begin by remembering the Sacred Space present – the sensuousness of the Present Moment, and our deep Desire for it.

Celebrant: All hail the East, presence of Water!
All: All hail the East, presence of Water!

---

[508] As Aphrodite is praised in the Orphic Hymns: http://www.asphodel-long.com/html/orphic_hymns.html.

Celebrant: We desire you, thirst for you, as all have done before us.
All: We desire you, thirst for you, as all have done before us.
Soft but constant drums, as co-celebrant takes water jug around, pours water into hands of each, each one drinks and responds:
I desire you, thirst for you, as all have done before me.

Celebrant: All hail the North, presence of Fire!
All: All hail the North, presence of Fire!
Celebrant: We desire you, gather round you, as all have done before us.
All: We desire you, gather round you, as all have done before us.
Soft but constant drums, as co-celebrant lights the fire.
Each comes to flame, passes their hands over and responds:
I desire you, gather round you (am drawn to you), as all have done before me.

Celebrant: All hail the West, presence of Earth!
All: All hail the West, presence of Earth!
Celebrant: We desire you, are held by you, desire to make our mark with you, as all have done before us.
All: We desire you, are held by you, desire to make our mark with you, as all have done before us.
Soft but constant drums, as co-celebrant attends the wet earth and towel, and each comes to the plate of wet earth, places their hand on it firmly and marks a piece of card paper, and responds:
I desire you, to make my mark with you, as all have done before me.
Each rinses their hand with assistance of attendant co-celebrant.

Celebrant: All hail the South, presence of Air!
All: All hail the South, presence of Air!
Celebrant: We desire you, reach for you, as all have done before us.
All: We desire you, reach for you, as all have done before us.
Soft but constant drums, as co-celebrant takes perfumed flowers around to each for them to breath in, and respond:
I desire you, reach for you, as all have done before me.

Celebrant: We are united in our Desire – for Water, Fire, Earth, and Air (PACING THE CIRCLE) – with each other, and with all who have come before us and all who will come after us, in their Desire and their

longing. Our desire, our longing, **is** the Beloved, is the Universe, desiring in us. Feel it – this Holy Desire. This is the Centre of Creativity – we are between the worlds, beyond the bounds of time and space, where light and dark, joy and sorrow, birth and death meet as One.
(Celebrant light centre candle)

Invocation
Celebrant: "Hear the words of the Star Goddess, the dust of whose feet are the hosts of heaven, ...

I who am the beauty of the green earth and the white moon among the stars and the mysteries of the waters, I call upon your soul to arise and come unto me. For I am the soul of nature that gives rise to the Universe. From me all things proceed and unto Me they must return. Let My worship be in the heart that rejoices, for behold – all acts of love and pleasure are My rituals. Let there be beauty and strength, power and compassion, honour and humility, mirth and reverence within you. And you who seek to know Me, know that your seeking and yearning will avail you not, unless you know the Mystery: for if that which you seek, you find not within yourself, you will never find it without. For I have been with you from the beginning, and I am that which is attained at the end of Desire."[509]

MUSIC ON (suggested: "Benediction Moon" by Pia, New World Music, 1998)
Celebrant: Name yourself as the Beauty whom She desires, the Beloved. Speak if you wish, of the beauty that you are, or simply show us, or tell us. Let us welcome your beauty.
Each one: (wording if and as they wish, presenting object or photo of Beauty, or describing, as they speak).
Suggested: "I am this beauty" AND/OR "I am the beauty of …"
Each put object or photo on the altar.
All respond: Welcome, we saw you coming from *afar*, and you were beautiful. We saw *you* coming from afar, and you ***are*** beautiful.
Each responds: It is so.

---

[509] Excerpt from *The Charge of the Goddess* by Doreen Valiente, in Starhawk, *The Spiral Dance*, 102-103.

Celebrant: Where there is perception of beauty there is love. With all this Holy Desirable Beauty let us proceed to the tree, the axis mundi, and dance the dance of life. Let us proceed contemplating this awesome beauty.

The Dance
(Instruction: Celebrant as #1, person next as #2. All 1's face right, all 2's face left. All 1's go in & under first, all 2's go out & over first.)
Celebrant: Take a ribbon of your choosing and weave into your life what you will, your heart's desires.
Each person: I choose red, for ... I choose blue, for ... etc
All respond: May it be so.

Chant – once before dance begins, then as the group dances:
"We are the Dance of the Earth, Moon and Sun
We are the Life that's in everyone
We are the Life that loves to live
We are the Love that lives to love."[510]

After it is woven, the chant and dancing continues, with drums. All move to firepits/fire cauldrons.

Leaping the Fire
Celebrant light the flames.
Celebrant: Let the flames of Love burn away your petty disharmonies and habitual negativities. Leap the Flames, or run between them, and leave behind what you will.

Each takes a turn saying what they leave behind to the flames and leaping the fire or running between the flames. Some may need to simply walk between the flames or pass their hands over them.

Drum roll for each.
All respond with claps and cheers and "May it be so."

---

[510] A slight variation of the chant written and taught to me by thea Gaia.

Sitting/Hands on Earth – when the excitement quietens down.
Celebrant: Let us sit on Mother Earth with this healing, and name others for whom we desire this.

After the naming, celebrant: Feel this desire in you now. This is the primordial essence, which brings forth all things. May all these desires spoken and unspoken, be so.

MOVE BACK INSIDE OR TO THE ALTAR. RE-LIGHT FIREPOTS
Celebrant: Let us walk again between the firepots, the eyes of Goddess, being *seen* by Her.

Communion
DISPLAY AND PARADE CAKE … mmm noises and comment.
Celebrant: Take and eat this holy cake, with its sacred honey – this Yoni-verse! ... consume your desire. *Consummate* your desire.
Pass the cake sunwise.
Each one repeat blessing to next: Take and eat this holy cake, consume your desire. *Consummate* your desire.
Server take glasses to all.

Celebrant holding up the juice: Take and drink this sweet nectar. Taste and enjoy the sweetness of life. Don't miss the Moment.
Pass sweet juice/wine sunwise around the circle.
Each person repeat the blessing: Taste and enjoy the sweetness of life. Don't miss the Moment.

Serve variety of berries and/or heart-shaped chocolates.

Stories (with ceremonial manners)
Let us tell each other stories of passion.
The group listens to each – it is not a conversation and a formal response is given to each.

A celebrant may recite Thomas Berry's poem *Earth's Desire.*[511]

Celebrant holding up bottle of champagne: This Dance of Life is an event, a party – Gaia's party, and we may be the champagne.[512] May we fulfill our role in existence – as a species … personally and collectively. May you enjoy a lightness of being.

Open the Circle
All turning to the South
Celebrant: We have remembered this evening the presence of Air, how we reach for Her. May we know peace within us.
All: May we know peace within us.

All turning to the West
Celebrant: We have remembered this evening the presence of Earth, how we are held by Her, how we desire to make our mark with Her. May we know peace within us.
All: May we know peace within us.

All turning to the North
Celebrant: We have remembered this evening the presence of Fire, how we are drawn to Her. May we know peace within us.
All: May we know peace within us.

All turning to the East
Celebrant: We have remembered this evening the presence of Water, how we thirst for Her. May we know peace within us.
All: May we know peace within us.

All turning to Centre
Celebrant: We have remembered that we are the Beauty whom She Desires. We have called this Beauty forth within ourselves and within

---

511 See Appendix H.
512 This is a suggestion of Brian Swimme's, *The Universe is a Green Dragon,* 123.

each other. We have remembered that the Holy Desire of the Universe is within us, uniting us with all and with the cosmos. We have woven into our lives our heart's desires, and received the Flames of Love. May we know peace within us and between us.
All: May we know peace within us and between us.

May the Delight and Ecstasy and Creativity of Earth, Moon and Sun, Beloved and Lover, She Who is All, go in our hearts and minds.
Pass the kiss

Song:
The circle is open but unbroken, the peace of the Goddess be ever in our hearts. Merry meet, and merry part, and merry meet again.
Blessed Be!

Music: suggested "Beltane Fires" by Loreena McKennitt

Points for individual contemplation prior to the ritual:
- your desires as the Universe desiring in you
- how you are the Beauty of the rose, the tree, or whatever you choose … "the Beauty whom She desires" Or "Wholly Desirable Beauty" whom *you* desire.
- what you would weave into your life
- what you would leave behind to the flames – let the Flames of Love burn away.

Beltaine/High Spring Goddess Slideshow:
https://thegirlgod.com/pagaianresources.php

CHAPTER 9
SUMMER SOLSTICE/LITHA

Southern Hemisphere – December 20-23
Northern Hemisphere – June 20-23

In this cosmology Summer Solstice is particularly a celebration of *She Who is this Dynamic Place of Being*, the Mother/Creator aspect of the Triple Goddess – as is the Winter Solstice (which is happening in the opposite Hemisphere at the same time – see diagrams in Chapter 2). This aspect of the Creative Triplicity is associated with the *communion* quality of Cosmogenesis.[513] The Young One/Virgin of Spring came into relationship with Other at Beltaine/High Spring and her face has changed into the Mother of Summer. This is the ripe fullness of Creativity. It is a Gateway – a Creative Interchange, like Winter Solstice, but this time the transition is from light back into dark. It is a *Birthing Place* – into larger self (whereas the Winter Gateway is a birthing place into unique differentiated being, into manifestation). Thus, I represent it with the Omega-yonic shape of the horseshoe: it is the full opening – the 'great O', the Omega. I take this inspiration from Barbara Walker's description of the horseshoe in her *Woman's Encyclopedia of Myths and Secrets*, as Goddess's symbol of "Great Gate;"[514] and her later connection of it with the Sheila-na-gig yoni display.[515]

At its deepest esoteric meaning, Summer Solstice is a celebration of the Mystery of Union – Re-Union – of this manifest form with All-That-Is, in its everyday reality. It is a celebration of the fullness of expression of this manifest form – the small selves that we are, and the paradoxical dynamic that as we fulfill our purpose, we dissolve.

Some Summer Solstice Story

The 'moment of grace'[516] that is Summer Solstice, marks the stillpoint in the height of Summer, when Earth's tilt causes the Sun to

513 This quality of communion is described by Swimme and Berry in *The Universe Story*, 77-78.
514 Walker, *The Woman's Encyclopedia of Myths and Secrets*, 414-415.
515 Walker, *The Woman's Encyclopedia of Myths and Secrets*, 931-932.
516 As Thomas Berry named the Seasonal transitions.

begin its 'decline': that is, its movement back to the North in the Southern Hemisphere, and back to the South in the Northern Hemisphere. This Seasonal Moment is polar opposite Winter Solstice when it is light that is "born," as it may be expressed. At the peak of Summer, in the bliss of expansion, it is the dark that is "born." Insofar as Winter Solstice is about birth, then Summer Solstice is about death. It is a celebration of profound mystical significance, that may be confronting in a culture where the dark is not valued for its creative telios.

Summer Solstice is a time for celebrating our realized Creativity, whose birth we celebrated at Winter Solstice, whose tenderness we dedicated ourselves to at Imbolc/Early Spring, whose certain presence and power we rejoiced in at Spring Equinox, whose fertile passion we danced for at Beltaine/High Spring. Now, at this seasonal point, as we celebrate light's fullness, we celebrate our own ripening – like that of the wheat, and the fruit. And like the wheat and the fruit, it is the Sun that is in us, that has ripened: the Sun is the Source of our every thought and action. The analogy is complete in that our everyday creativity – our everyday actions, and we, ultimately, are also "Food for the Universe"[517] … it is all how we feed the Universe.

Like the Sun and the wheat and the fruit, we find the purpose of our Creativity in the releasing of it; just as our breath must be released for its purpose of life. The symbolism used to express this in ceremony has been the giving of a full rose/flower to the flames.[518] We, and our everyday creativity, are the "Bread of Life," as it may be expressed; just as many other indigenous traditions recognize everyday acts as evoking "the ongoing creation of the cosmos,"[519] so in this tradition, Summer is the time for particularly celebrating that. Our everyday lives, moment to moment, are built on the fabric of the work/creativity of the ancestors and ancient creatures that went before us; and so the future is built on ours. We are constantly consuming the work and creativity of others and we are constantly being consumed. The question may be asked: "Who

---

[517] Swimme uses this expression in *Canticle to the Cosmos,* video 5 "Destruction and Loss."

[518] This is based on the traditional Litha (Summer Solstice) rite described by Starhawk, *The Spiral Dance*, 206.

[519] Spretnak, *States of Grace*, 95.

are you feeding?,"[520] and consideration given to whether you are happy with the answer.

We celebrate the blossoming of all Creativity then, and the bliss of it, at a time when Earth may be pouring forth Her abundance, giving it away. We aspire to follow Her example. In this cosmology, what is given is the self fully realized and celebrated, not a self that is abnegated – just as the fruit gives its full self. Everyday tasks can be joyful, if graciously received, and not rushed by: I think of Eastern European women I have seen singing as they work in the fields, in the Marija Gimbutas film *Signs Out of Time* – it may be a good thing to practice, if it has been forgotten.

Summer Solstice is a celebration of the *Fullness of the Mother* – in ourselves, in Earth, in the Cosmos. We are the Sun, coming to fullness in its creative engagement with Earth. We affirm this in ceremony with: "It is the Sun that is in you, see how you shine." It is the ripening of Her manifestation, which fulfills itself in the awesome act of dissolution. This is the mystery of the Moment. Brian Swimme has described this mystery of radiance as a Power of the Universe, as *Radiance*: the shining forth of the self is at the same time a give-away, a decline of the self – just as the Sun is constantly giving itself away.

This Solstice Moment of Summer is a celebration of communion, the feast of life – which is for the enjoying, not for the holding onto. Summer and Winter Solstices are Gateways – between the manifest and the manifesting, and Summer Solstice is a Union/Re-Union of these, a kind of meeting with the deeper self. Winter Solstice may be more of a separation, though it is usually experienced as joyful, because it is also a meeting, as the new is being brought forth. The interchange of Summer Solstice may be experienced as an entry into loss – the Cosmological Dynamic of Loss, as manifestation passes. Beltaine, Summer Solstice and Lammas – the next Seasonal Moment, may be felt as the three faces of Cosmogenesis in the movement towards entropy.[521] The light part of the

---

[520] As Swimme asks in *Canticle to the Cosmos*, video 5 "Destruction and Loss."

[521] Just as Samhain, Winter Solstice and Imbolc may be felt as the three faces of Cosmogenesis in the movement towards toward form – syntropy.

annual cycle of Earth around Sun is a celebration of the Young One/Virgin quality of Cosmogenesis, with Her face gradually changing to the Mother/Communion quality; and through the Autumn, the dark part of the annual cycle, it is a celebration of the Old One/Crone quality, whose face will gradually change also, back to the Mother/Communion. They are never separate.

In this cosmology desire for full creativity has been celebrated as the allurement of the Cosmos, and being experienced as gravity, as relationship with Earth, our place of being, how She holds us. At both Solstices there is celebration of deep engagement, communion.

With the peaking of light at Summer Solstice, many of the fruits of Earth are coming to their fullness, many grains ripen, deciduous trees peak in their greenery, lots of bugs and creatures are bursting with business and creativity. We may celebrate our own ripening, the maturing of creativity, the bliss of our full expression – in every moment, as well as in special achievements, and ultimately. In the Early Spring/Imbolc it is the Child of Promise[522] in each and all that is remembered, and the tenderness and vulnerability of our being that is nurtured. At this season of Summer Solstice we may celebrate that Promise growing to fullness in us, being received and given – expressed as the *Sacred Fullness of the Mother.*

The Beauty that we dedicated ourselves to at Imbolc/Early Spring (the differentiated self), that we identified with at Beltaine (I am the Beauty) – is now given back, given away, ripe for consuming. This is its purpose, its sacred destiny. We do desire to be received, to be consumed – it is our joy and our grief. Summer then is like the rose, as it is said in Pagan tradition[523] – with blossom and thorn … beautiful, fragrant, full – yet it comes with thorns that open the skin. All is given over.

In traditional versions of Paganism, the story is that Beloved and Lover (often storied as Goddess and God, though there is no need), reach the full expression of their Love at this time – the Allurement of Beltaine/High Spring, finds its maturation. Beloved and Lover could also be understood as *deep creative self and small self*, and in their union the

---

522 An expression used by Starhawk in *The Spiral Dance.*

523 Starhawk, *The Spiral Dance*, 205.

boundaries of the self are broken, they merge. Beloved and Lover embrace, in a Love "so complete, that all dissolves into a single song of ecstasy that moves the worlds."[524] It may be understood as the *Big Orgasm* – the "little death" as it is known in some cultures – the realization of Union with the Beloved. The Old story tells us then that this ecstasy of union is at the core of the mystery of existence – it is the fullness and end of Desire. All is given away – all is poured forth, the deep rich dark stream of Life flows out. In PaGaian Summer ceremony there is affirmation that each of us is "Gift"; and that is understood to mean that we are both *given and received* – all at the same time. The breath is given and life is received. We receive the Gift with each breath in, and we are the Gift with each breath out.

By way of note, in Pagan traditions of recent times it has been common to story the Sun, our Star, as male: this is unnecessary and may not even be an appropriate metaphor. There is plenty of precedent for Sun being storied as Mother/female in cultures around the world. Patricia Monaghan develops this appropriation of Goddess capacities and the embedded duality from which it springs, in her book *O Mother Sun: a New View of the Cosmic Feminine.* The title of the Introduction is "The Apollo Conspiracy."[525]

Summer Solstice/Litha Meditation

To begin this Seasonal Moment's contemplation:

> Centering for a moment … feeling the full cycle of your breath, and recognising the Triple Goddess Dynamic here in this primary place: She who unfolds the Cosmos … receiving the Gift, as you inhale – feeling the Urge to Be, feeling the peaking – the exchange, the communion that this Place of Being is, then the release – the dissolving, becoming the Gift, making space, ever transforming. Knowing the Triple Spiral dynamic here in your breath, the Wheel of the Year in your breath.

---

524 Ibid.

525 1-7.

Some Summer Solstice Motifs

[Figure 24] A Front Door for Summer Solstice

Suggested decorations for this Seasonal Moment – on the front door, windowsills, corners of shelves, and/or altars – may include:

- **ripe fruit, bread, grain** – representing fulfilment of purpose, fulfilment of creativity.
- **full flower, and full green leaves** – representing fullness of being.
- **yellow fabric disc** – representing full sun and light.
- **black ribbon** – representing the dark that is born.

An Invocation

Then the Eye of Certainty opens, and staring inwardly at herself, the lover finds herself lost, vanished. But ... she finds the Beloved; and when she looks still deeper, realizes the Beloved is herself. She exclaims,

"Beloved, I sought you
here and there,

asked for news of you
from all I met;
then saw you through myself
and found we were identical.
Now I blush to think I ever
searched for signs of you."[526]

The Universe Story and Summer Solstice

A supernova event may be understood as expressing the poiesis of Summer Solstice: the Star matures, after billions of years of shining forth, and explodes. It is "simultaneously a profound destruction and yet an exuberant creativity."[527] As Swimme and Berry tell it:

> The more closely we look at any place in the fifteen billion years of the universe's story, the more we realize that the universe is both violent and creative, both destructive and cooperative. The mystery is that both extremes are found together. … (this is because) our universe is self-energising. All the energy of the universe is needed by the entire universe for its own development. No development anywhere in the Universe can take place without energy. The second law of thermodynamics points out that constructive activity needs energy and inevitably produces entropy, or waste. Every development has a cost, an inescapable cost. A price must be paid for creativity; an energy payment must be made for maintaining beauty; an energy payment is demanded for any and all advance.[528]

There is a *sacrificial* dimension that the universe has. There is difficulty with the word *sacrifice* because of a history of misuse, but "for our ancestors a sacrificial act was a way of making holy."[529] A supernova

---

526 From "Divine Flashes" by Sufi poet Iraqi, Chittick and Wilson, 76.
527 Swimme and Berry, *The Universe Story*, 49.
528 Ibid., 51-53.
529 Ibid., 59.

explosion is a sacrificial event, that "shapes the future and has been prepared for since the beginning of time."[530]

Brian Swimme develops this into how we (all beings) are Food for the Universe. He says: "Every moment of our lives disappears into the ongoing story of the Universe. Our creativity is energising the whole:"[531] thus we may understand ourselves as "the Bread of Life." In the traditional Pagan version of the story, the God is the Bread – the grain that has ripened and become food.[532] But in the earliest stories, the Goddess Herself was the grain:[533] and I interpret that to mean that we are each the grain … the Sun ripens in all of us, we are each required to become food, we *are* food whether we like it or not. The Mystery of the Universe is present in each and all.

Individual creativity is given away, and the model for this "Give-Away" as Starhawk names it,[534] may be the generosity of the Sun, which is the very source of our sustenance. Starhawk calls the Summer Solstice the "Give-Away time of the Sun" and expresses: "she feeds us from Her own body."[535] It is the Sun in us that shines.

This cosmology assents to and nurtures a concentration of Being that innately demands to be poured forth: it creates a generosity within, since abundance is its very nature. We celebrate the innate generosity of the Universe, for which Sun may be our model.[536] Summer Solstice may be a remembering that "There is not a single solitary thought or action in the history of humanity that is not a Solar event."[537] It is a time for

---

530 Ibid., 49.

531 *Canticle to the Cosmos*, video 5 "Destruction and Loss."

532 Starhawk, *The Spiral Dance,* 205-207. Starhawk also discusses "whether this figure should be specifically male, androgynous, male and female both, or an abstract representation of the sun," 260.

533 See Berger, *The Goddess Obscured.*

534 *The Spiral Dance,* 236.

535 Ibid.

536 See Brian Swimme, "The Generosity of the Sun" in *The Hidden Heart of the Cosmos* video, Center for the Story of the Universe: program 1. See https://archive.org/details/hidden-heart-of-the-cosmos-1996.

537 Swimme, *Canticle of the Cosmos*, video 2 "The Primeval Fireball."

remembering our Source, and the ongoing Event that we are part of, that even the Sun itself is part of.

And a quote from *The Universe Story,* that is resonant with Summer Solstice:

> In the primeval fireball, which quickly billowed in every direction, we see a metaphor for the infinite striving of the sentient being. An unbridled playing out of this cosmic tendency would lead to ultimate dispersion. But the fireball discovered a basic obstacle to its movements, the gravitational attraction. Only because expansion met the obstacle of gravitation did the galaxies come forth. In a similar way the wings of birds and the musculature of the elephants arose out of the careful embrace of the negative or obstructing aspects of the gravitational attraction. Any life forms that might awake in a world without gravity's hindrances to motion would be incapable of inventing the anatomy of the cheetah.[538]

Notes on imagery and metaphors in the offered Ceremonial Script

This Season is about fullness of expression – being an "open channel for the moving energies of Life," as is expressed in the offered ceremonial script; so it is good if the participants come into the circle moving to the rhythm of the drums and/or other music.

The focus of the Centering Breath Meditation is on the experience of the peaking/fullness of the breath, beginning with mindfulness of the experience of being held by Earth, and the connection of this experience to the power that holds Earth in orbit around Sun, and Moon around Earth: it is meant as a sensing of everyday full relationship with cosmic power, how it is always present, usually simply below consciousness. There is also a "banishing process" (after the Statement of Purpose) which would traditionally be understood as "cleansing the space": it is expressed as making conscious those inner voices or energies that impede one's full expression and creativity.

For Calling the Directions/Elements – "Casting the Circle" the focus repeats that of Winter Solstice, recalling the elements as Cosmic Dynamics present in human capacities, and thus reflecting a seamless Gaian self; however at this Summer Moment there may be specific

---

[538] Swimme and Berry, *The Universe Story,* 55.

invitation for each one to speak of or be conscious of their relationship with each element as it is brought to them. This is an emphasis on the fullness of expression, as well as on the Cosmogenetic quality of relatedness and deep communion.

The Invocation is a recognition and drawing forth – in self and other – of the fullness of being, the wholeness of the Universe, and the radiance of Sun our source of life and energy. Each is blessed with "The Summerland is within you": for the ancients of Old European Pagan tradition, "Summerland" was a place similar to "heaven" of the Christian mind, but there was nothing "pie in the sky" about it. It was a place of eternal fulfillment and immediately present. Some have understood "Avalon" in this way. "Summerland" as I mean it here in the offered script is a Place where we may dwell now, even in the midst of awareness of the awesomeness of life, where we know and sense that we are immersed in the flow and richness of the Universe. I mean it primarily as a state of being wherein we act with ease and pour forth our essence: a state of fullness of expression of deep self, which I understand as a state of a sense of sacred union with Earth, Sun, Moon, Cosmos, Creativity … however this Cosmic harmony and graceful ease of being may manifest within each.

The invocation is continued with a "Dyad Poem" or "Conversation of Union" as I call it (from Starhawk's *The Spiral Dance*),[539] wherein at this Solstice gateway from the light into the dark, an inner and an outer circle of participants express the reciprocity and union of form and formlessness. This dialogue poem may express the flux and interchange of the "manifest" and the "manifesting" of the Cosmos. This formation of an inner and an outer circle in conversation also recalls the formation at the Winter Solstice invocation for the Cosmogenesis Dance – which is done in silence. The Summer Solstice dyadic conversation/poem culminates in a unified prayer/invocation and then a toning – a harmonising of voices that may be felt as an exercise in relationship with self, other and all-that-is … the three layers of Cosmogenesis: that is, each is feeling for their own true expressive voice, listening for the others, and it is all coming out of and going into a larger

---

[539] This "Conversation of Union"/"Dyad Poem" is an adaptation of Starhawk's "Invocation to the Ground of Being," *The Spiral Dance*, 131.

self. The toning that arises is co-created and yet it is coming from the Well of Creativity itself. It may be a magical experience.

The Spiral Dance done in the Summer Solstice ceremony is a reflection of the Spiral Dance in the Winter Solstice ceremony wherein it was done with lit candles in hand, representing the small flames being born in each individual. Now at Summer Solstice, the spiral is done with the full rose/flower in hand, signifying the full radiance of what each may/have become, and which is given back to Source.

The rite of giving the full rose/flower to the fire as is done may be understood in multivalent ways, and each will have their own subtle understandings. Essentially it is to signify the giving away of the fullness of our creative selves (which is counter to the "birth" of this creative self at Winter Solstice when candles are lit to signify that). Throwing the rose/flower on the fire may be seen as a commitment to follow your passion, and to release it, express it – to feed the world with your fullness of being, the fullness of your particular self, wherein the Universe speaks. This tradition is about fulfillment of the self – giving that to the Universe, not self-abnegation. Note that this is not about burning away negatives in yourself (as is popular to do); it is about giving your full potential and becoming that – in every moment, and ultimately.

The ceremonial rite with the wreath being upheld for each participant to look through to the fire, being admonished and encouraged to see with clear sight, is from Starhawk's tradition.[540] It is an invocation of clarity of vision – of birth and death, the "Om" and the "Omega." When it is held aloft for the individual to see the circle of full roses, it is an affirmation of the unbroken circle: that is, an affirmation of the Cosmic Creative Spin as never-ending. Nothing is wasted – all comes full circle.

Since Summer Solstice Moment is quintessentially a celebration of the quality of communion, for the Communion rite each participant takes the ceremonial loaf of bread in their own hands and breaks off a piece, affirming that they are the ripened grain, "the Bread of Life." Each thus may express the ripening in themselves of the *Promise* of the light part of the cycle, whose purpose is to be given to Other and All-That-Is and consumed. Each one similarly affirms this in the consuming of the

---

[540] *The Spiral Dance,* 206.

wine/juice. It also affirms that it is the Sun in each that has ripened, that "we are the Sun."

In the Storytelling space after Communion, there is a suggestion to "boast" as it seems an appropriate Summer Solstice thing to do: it is good practice for many to speak well of themselves and their successes.

Throughout the Summer Solstice ceremony in general there is a mirroring of the Winter Solstice ceremony; it is to affirm their connection as places of interchange, celebrating the Mother/communion quality in particular, one the Om and the other the Omega.

[Figure 25] A Summer Solstice/Litha Ceremonial Altar

A SUMMER SOLSTICE/LITHA CEREMONIAL SCRIPT

Note that the directions here are called in "counter-clockwise" direction which is sunwise in the Southern Hemisphere. The order may be changed to "clockwise"; sunwise for the Northern Hemisphere. Note that this particular offered ceremonial script is not considered prescriptive, and that there are many creative possible variations.

Some requirements and some suggestions: light yellow altar cloth, with black ribbon. Fire ready off to the side for burning the flowers. Each brings a full rose/flower of choice (not a bud). Wrapped lollies (candy), or in small bags. Baskets of summer fruits – cherries, plums, apricots, nectarines, grapes, berries, and baskets of small blossoms. Wreath of roses, and/or wildflowers, with black ribbon tied in. A loaf of good bread (round is good) near fireplace and another on the altar. Herbs for the fire. Bowl of scented oil (rose is good). White wine, grape juice (white is good for this Season). Paper bags for each for fruit & lollies. Copies of Dyad Poem for each. Olive oil in small jugs. Small bowls for dipping bread in oil for each. Rattles, tambourines, drums. I also like a large image of Inanna displaying her breasts – FOOD – the primary and eternal communion experience (see Summer Solstice Goddess Slideshow linked at the end of this chapter).

Each one take their rose/flower to pre-ritual contemplation with them; place in basket on altar upon gathering to the circle.

Call to Gather … drumming. As people come in, moving around the circle feeling the rhythm.

Centering – Breath Meditation

Breathe deep, and sink into the weight of your body – feel Earth holding you. This is how She holds you – feel your weight, Earth holding you. With this same power Earth is held in orbit around Sun, Moon around Earth. Feel this Power in you … this relationship with Earth, Sun and Moon – this Cosmos. Breathe into it.

Breathe deep, and draw up from the Mother all that you desire … filling your bodymind to capacity, until you can draw it no longer, feeling the fullness at the peaking of your breath, and the urge to release it … then letting it go. Again, breathing deep, filling your bodymind to capacity – feeling the Fullness at the peaking of your breath – holding it a moment, and feeling the urge to let it go, to pour it forth – then letting it go … feeling this Sacred Interchange of form and formlessness – the fullness of being and the pouring forth of it.

This is how Gaia breathes in our part of the world at this time. She is filling to capacity, and giving it away, letting it go. Joining with Her now in this breath … breathing deep, filling to capacity, feeling Her Fullness and the urge to release, let go, pour it forth – and as you do, recalling those times in your life when have acted with ease and poured forth your essence. This is the Summerland within you … the Sacred Interchange of form – the fullness of being, and formlessness – the dissolving. Feel it in your breath.

Statement of Purpose

This is the time of the Summer Solstice in our Hemisphere – when the light part of the day is longest. In our part of the world, light is in Her fullness, She spreads Her radiance, Her fruits ripen, Her greenery is peaking, the cicadas sing. Yet as Light reaches Her peak, our closest contact with the Sun, She opens completely, and the seed of darkness is born.

As it says in the tradition, this is the season of the rose, blossom and thorn, fragrance and blood. The story of Old tells that on this day Beloved and Lover embrace, in a love so complete, that all dissolves, into the single Song of ecstasy that moves the worlds. Our bliss, fully matured, given over, feeds the Universe and turns the wheel. We join the Beloved and Lover, Earth, Moon and Sun, in the Great Give-Away of our Creativity, our Fullness of Being.

Banishing – Personal and Planetary dimensions

Let us begin by remembering that we are each and all open channels for the moving energies of life – for the bliss of union of Beloved and Lover, the Creativity of Earth, Moon and Sun.

Let us raise some energy to banish those blocks that hold us back from being these open channels, fully who we might be – raise some energy for healing/wholing as many of our ancestors did of Old in this season.[541]

---

[541] See *The Midsummer Dancers* by Max Dashu https://www.suppressedhistories.net/secrethistory/dancers.html. It is a chapter from her unpublished book *Female Spheres of Power*, Vol XII in her series *Secret History of the Witches*.

DRUMS AND RATTLES AND VOICE AND WHIRLING.
Chant: *Mother Earth who gives us birth: You feed all, and all feed You.*[542]

Having raised energy to banish those states of being which block our flow, let us call upon those with which we may be in harmony.

Calling the Directions/Casting the Circle
All turning to the East
Celebrant: Hail the East, Powers of Water, Cosmic Dynamic of Sensitivity.[543]
All: Hail the East, Powers of Water, Cosmic Dynamic of Sensitivity.
Celebrant: you absorb, become, whatever you touch; may we *feel* what we are, and respond compassionately.
All: may we *feel* what we are and respond compassionately.

DRUMS as co-celebrant sprinkles water with oak or native branch (each may speak as they wish as water goes by, conscious of personal relationship with water).

All turning to the North
Celebrant: Hail the North, Powers of Fire, Unseen Shaping Power of the Cosmos.[544]
All: Hail the North, Powers of Fire, Unseen Shaping Power of the Cosmos.
Celebrant: that gives us form – flames that we are; may we dance with you and act with Creative Lust for all of life.
All: May we dance with you and act with Creative Lust for all of life.

DRUMS as co-celebrant lights fire in the pot and carries it around (each may speak as they wish as fire goes by, conscious of personal relationship with fire).

---

[542] I am thankful to Erin Parkinson for this chant (Summer Solstice 2015).
[543] This understanding of Water comes from Swimme, *The Universe is a Green Dragon*, 87-95.
[544] This understanding of Fire comes from Swimme, Ibid., 127-139.

All turning to the West
Celebrant: Hail the West, Powers of Earth, deep Sentient Presence and Memory.[545]
All: Hail the West, Powers of Earth, deep Sentient Presence and Memory.
Celebrant: You hold all the stories of life in your Body – as each of our bodies do also; may we remember who we really are, may we hold the wisdom of all time and no time.
All: may we remember who we really are, may we hold the wisdom of all time and no time.

DRUMS as co-celebrant holds up rock/bowl of earth and carries it around (each may speak as they wish as earth goes by, conscious of personal relationship with earth).

All turning to the South
Celebrant: Hail the South, Powers of Air, Cosmic Dynamic of Exuberance and Expression.[546]
All: Hail the South, Powers of Air, Cosmic Dynamic of Exuberance and Expression.
Celebrant: Wind that moves the trees, the clouds, brings us rain and allows us our voice; move us and inspire us to unfurl our being.
All: Move us and inspire us to unfurl our being.

DRUMS as co-celebrant lights smudge stick and carries it around (each may speak as they wish as smudge goes by, conscious of personal relationship with air).

All turning to the centre
Celebrant: This is what we are ... mysterious flashes of Water, Fire, Earth and Air (PACING THE CIRCLE) – in this Dynamic Place of Being, this Sacred Interchange of form and formlessness – this Event. We are at the Centre. The circle is cast, we are between the worlds, beyond the bounds of time and space, where light and dark, birth and death, joy and sorrow

---

545 This understanding of Earth comes from Swimme, Ibid., 99-109.
546 This understanding of Air comes from Swimme, Ibid., 143-151.

meet, as one.
(Celebrant light centre candle)

Invocation
Celebrant: "Let us recognize and invoke the sacred Fullness of the Mother, the Wholeness of the Universe, the Radiance of Sun – the Summerland, within us and in each other."

"Take the bowl of oil when it comes to you and turn to the person next to you, anoint their hands or forehead, bless them in this way:
Hail ...., thou art the Sacred Fullness of the Mother, the Wholeness of the Universe: the Radiance of Sun, the Summerland, is within you, and bow deeply as you recognize all that is before you."

Celebrant begin with person on right (going sunwise): then each person pass the invocation to the next, around the circle

Response of each: It is so (and whatever else they may like to affirm).

Celebrant: "Let us continue this invocation of the sacred Wholeness present in us with a Conversation of Union – of the form and formlessness that we are":
Dyad Poem[547] (inner and outer circles – facing each other)
CHANT – each circle repeating each phrase 3 times

Inner:	**Outer:**
Nameless One	of many names
Eternal	and ever changing One
Who is found nowhere	but appears everywhere
Beyond	and within all
Timeless	circle of the seasons
Unknowable Mystery	Known by all
Mother of all Life	Young One of the Dance
Engulf us with your love	Be radiant within us

547 See Starhawk, *The Spiral Dance*, 131, as noted earlier.

Inner circle step back into outer circle:
All (with actions): See with our eyes. Hear with our ears. Breathe with our nostrils. Kiss with our lips. Touch with our hands. Open our hearts. That we may live free … joyful in the Song of all that is.

Song/Toning – the Uni-Verse
Each begin with their own sound, building on this, harmonising voices. Let it build and come to stillness in its own time.

The Give-Away
Co-celebrants (two or three), after a period of stillness: throw light blossoms in the air over the participants, with the blessing
She gives it away, She pours it forth.
Then hand out small bags of sweets, then seasonal fruits, with:
She gives it away, She pours it forth.
(Everyone puts the sweets and fruits in their paper bags)

Spiral Dance
Celebrant: "Each pick up your full flower or rose, which represents your Fullness of Being, your radiance – what you have drawn from the Mother and what you pour forth – and let us dance the mystery of the Spiral."

All pick up their rose/flower and take hands except for the leader of the spiral, holding their rose/flower in their linked right hand (moving sunwise to start).

All chant:
She is shining, crowned with light.
We are radiant, we are bright.
We dissolve into the night.[548]

As the circle opens back out, lead to the fireplace. Co-celebrant light fire.

Celebrant: Let us each join in the Great Give-Away, of our Fullness of Being, our Radiance, our Creativity, to the Universe. Give your full

---

548 My own variation of a chant in Starhawk, *The Spiral Dance,* 206.

flower to the flames, and speak if you wish, perhaps of what you pour forth every day, perhaps what has grown in you this year, or what you would *like* to pour forth, however you'd like to express it, if you would.

Celebrant throw some herbs on the fire – wormwood, sage or herb of choice.
Each one comes forward, and throws their full rose/full flower on the fire, speaking if they wish, and as they wish.
All respond: We bless you, and the Gift that you are.

Celebrant throws more herbs on the fire when all are done, and pronounces: "Long live the dance and those who are in it … even the stars will join in. May we be frisky, vibrant and lively!"[549]

Wreath[550]
A celebrant takes up the wreath. Each may come forward to the wreath held up in front of fire, or it may be taken around to each person.
Wreath is held so each can see through it to the flames, as celebrant says: "See with clear sight."
The celebrant then holds the wreath up in front of the person saying "And know the mystery of the unbroken circle."

Chant by group after all are blessed with the wreath:

One thing becomes another,
In the Mother, in the Mother.[551]

Communion
Celebrant holds up the loaf of bread which is near the fireplace, saying: "We are each the Bread of Life – feeding the world with our everyday

---

[549] A quote in part from and inspired by Max Dashu in *The Midsummer Dancers*
https://www.suppressedhistories.net/secrethistory/dancers.html.
[550] A process from Starhawk, *The Spiral Dance*, 206.
[551] I don't remember the reference for this; it may be Starhawk.

acts and being, ripening for consuming. It is the Sun that ripens in us – bringing us to this creative fullness, this Radiance. Step forward one at a time, take the bread in your hands, break off a piece and affirm that this is so – you are the Bread of Life: it is the Sun that ripens in you, you are the Sun ... however you may like to express it if you would."
Each comes forward in turn, takes the loaf of bread in their hands, breaks some off for themselves, and speaks an affirmation if they wish.

When all are done, celebrant affirms: "Let us eat. We are the Sun, we are the grain, we are the Bread of Life. We are Her Food."

All move back to centre altar/circle
Celebrant holding up the wine/juice says:
"We are each the Wine, poured out for the Mystery, coming to full flavour. It is the Sun that ripens in us – bringing us to this creative fullness, this full flavour. Step forward one at a time, take the wine in your hands, pour some out and affirm that this is so – you are the Wine: it is the Sun that ripens in you, you are the Sun ... however you would like to express it if you would."

Each comes forward in turn, takes the decanter and glass in their hands, pours some wine/juice for themselves, and speaks an affirmation if they wish.

When all are done, celebrant affirms: "Let us drink. We are the Sun, we are the grape, we are the Wine. We are Her Food."

Storytelling and "boasting" perhaps
Let us enjoy some of our fruits and lollies (candy) and tell each other stories of the fullness of Creativity, of Wholeness, what you are creating – perhaps every day. You may take this opportunity to boast about what you have created/achieved, about your radiance.

The group listens to each with ceremonial manners: that is, it is not a conversation, with only a formal response to each.
Response: "May you be fruitful." or "We bless your creativity …" as seems formally appropriate.

PASS AROUND SMALL JUGS OF OLIVE OIL for bread, and small dishes for each person. All may eat their sweets and fruits.

Open the Circle
All turning to the South
Celebrant: We have remembered this evening that we are Air
cosmic dynamics of exuberance, unfurling our being. May there be peace within us.
All: May there be peace within us.

All turning to the West
Celebrant: We have remembered this evening that we are Earth
deep sentient presence and memory. We hold all the stories of life in our bodies. May there be peace within us.
All: May there be peace within us.

All turning to the North
Celebrant: We have remembered this evening that we are Fire, vessels of unseen shaping power, dancing flames. May there be peace within us.
All: May there be peace within us.

All turning to the East
Celebrant: We have remembered this evening that we are Water, cosmic dynamics of sensitivity, absorbing, becoming all that we touch. May there be peace within us.
All: May there be peace within us.

All turning to the centre
Celebrant: We have remembered that we are each open channels for the moving energies of life – the Sacred Fullness of the Mother, the Wholeness of the Universe, the Radiance of Sun; that the Summerland is within us – that the Creativity that pours forth from us is Divine, Sacred – is what the Cosmos is made of. We have remembered that it is the Sun that ripens in us, we are the Bread of Life – that our Passion released, may feed the world. May there be peace within us and between us.
All: May there be peace within us and between us.

The circle is open but unbroken.
May the peace and bliss of the One, of union of Beloved and Lover, of Earth, Moon and Sun, Goddess, go in our hearts and minds. May we be radiant.

Song:
The circle is open but unbroken, the peace of the Goddess be ever in our hearts. Merry meet, and merry part, and merry meet again.
Blessed Be!
Music: I like to use Wendy Rule's "Radiate" from her *Guided by Venus* CD (2010).

Points for individual contemplation prior to the ritual:
- What holds you back from being fully who you might be – what blocks the flow?
- How you and others are the "sacred Fullness of the Mother, the Wholeness of the Universe," how the radiance of Sun, the Summerland, is within you.
- Contemplate the Dyad Poem as a conversation of the interchange of manifest form and 'manifesting' formlessness – always present.
- The full rose/flower – what you pour forth or have poured forth, to the Universe, how you feed the Universe with your daily creativity, your passion, your actions. It represents your fullness of being, your radiance – perhaps your achievements, what has grown in you this year or over a longer period.
- How you are the Bread of Life, how it is Sun that ripens in you, how everything you do and think is a Solar event: how you are Sun, and may feed the world just like Sun does – with your generous radiance.

Summer Solstice/Litha Goddess Slideshow:
https://thegirlgod.com/pagaianresources.php

CHAPTER 10
LAMMAS/LATE SUMMER

Southern Hemisphere – Feb. $1^{st}/2^{nd}$,
Northern Hemisphere – August $1^{st}/2^{nd}$
These dates are traditional, though the actual astronomical date varies. It is the meridian point or cross-quarter day between Summer Solstice and Autumn Equinox, thus actually a little later in early February for S.H., and early August for N.H., respectively.

In this cosmology Lammas/Late Summer is the quintessential celebration of the Old Dark Wise One, She who dissolves us – all – in every moment. I name Her as *She Who Creates the Space to Be*; She is the Old One/Crone face of the Triple Goddess, who is the deep sentience/subjectivity within all. This aspect of the Creative Triplicity is associated with the *autopoiesis* quality of Cosmogenesis,[552] the creative dark capacity of all being, directly participating in the Well of Creativity, to whom we all return. This Season of Lammas/Late Summer is an opportunity to identify with this quality and face within: we do, each and all, embody the transformations of the ages, we do emerge from this Larger Self and we do return to Her. This *identification* with the Old Dark One mirrors its opposite on the Wheel of the Year, the Imbolc/Early Spring celebrations wherein we identified with the new Young One, the *Urge to Be*, and claimed that embodiment (see diagrams in Chapter 2). This Lammas Seasonal Moment's identification with the ancient Dark One is an opportunity to remember our innate belonging and deep love of this Larger Self that we are always. The rest of the dark part of the cycle celebrates Her *processes*: of grief, the gaining of wisdom and deep transformation.[553] At this Moment, after the climax of Summer, all begins the dissolution back into Her, into Larger Self: all is/are given over. We may understand ourselves as embodying the harvest: that *we* are the harvest. It is a funerary moment, "the Sacred Consuming" as I name it. And whereas the old dominating dark space of Samhain/Deep Autumn is about "re-solution," the new darkening space of Lammas is about "dis-

---

552 Swimme and Berry, *The Universe Story*, 75-77.
553 See Chapter 2, "In Summary: Contemplating How Creativity Proceeds."

solution."

The Young One/Virgin of Spring and the light part of the cycle came into relationship with Other and her face changed into the Mother of Summer, which is now passing into the Old One. In the poietic process of the Seasonal Moments of High Spring/Beltaine, Summer Solstice and Lammas/Late Summer, one may get a sense of these three in a movement towards entropy, towards the manifesting realm: from the *differentiated* fertility and allurement of Beltaine, through the gateway and *communion* of Summer Solstice to the *autopoietic* dissolution/expansion of Lammas/Late Summer.[554] The three are a kaleidoscope, seamlessly connected. If one may observe Sun's position on the horizon as She rises, the connection of the three can be noted there also: that is, Sun at Beltaine/High Spring and Lammas/Late Summer rises in the same position for Beltaine and for Lammas, halfway between Summer Solstice and Equinox, but the movement is just different in direction.[555]

The dark is waxing; it is not just light waning. Late Summer/Lammas is a recognition of the *autopoietic* space aspect of Creativity, of Cosmogenesis; and we may recognise each being's direct participation in the Creative Cosmos. This *autopoietic* quality is celebrated also at Samhain/Deep Autumn, but now in different relationship with the other two qualities of *differentiation* and *communion* at this point around the Wheel: that is, at Samhain, this quality may be felt more as the beginning of the three, now at Lammas it may be felt more as the end of the three qualities, though the beginning and the end are never clearly distinguishable, they are always deeply involved with each other.

Of note here is a heretofore common dualistic and patriarchal association of the virgin with death (as temptation) that may in fact have some basis in this cosmology only from a very different perspective: that is, the Young One/Virgin of Beltaine waxes to maturity, peaks at the Summer Solstice and dissolves into the Old One/Crone in this part of

---

[554] This poietic process mirrors the one articulated in Chapter 6, of Samhain/Deep Autumn, Winter Solstice and Imbolc/Early Spring in a movement towards manifest form – syntropy.

[555] It is a very informative and simple process to note this in your place, to make markers of some kind, and observe it over the period of the year.

the cycle. These three Seasonal Moments participate in the same process of entropy as mentioned earlier, so they may be understood as "fuzzy,"[556] not simply linear, but all three in each other ... this is something recognised of Old, thus the Nine Muses, or the numinosity of any multiple of three.

Also of note at this point around the Wheel, is contemplation of other connections of the Seasonal Moments. One that I noticed is the triangular relationship of Samhain, Imbolc and Summer Solstice, overlapping with the opposite pointing triangular relationship of Beltaine, Lammas and Winter Solstice, which forms a six-pointed star. I named the emergent image as *The Star of Aphrodite.*[557]

Some Lammas/Late Summer Story

In this Season of the waxing dark, Earth's tilt takes this region back away from Sun. After the great act of radiance of the Summer Solstice when all is poured forth, given over, we dissolve into the night: the dark is born. This is the first harvest celebration of the Wheel's cycle, a celebration of the Dark as the holy realm into which we are dissolved. Like its polar opposite Imbolc/Early Spring, Lammas too is a dedication – this time to the Deep, the Old One within, Who returns us to Source, creates the Space to Be. Mortality is a transformation; we are consumed – harvested – so that Gaian life/Creativity may go on, that She may be ever-new. And it is the *native place* of all being, though it is common in

---

556 "Fuzziness" is a term used by scientist and philosopher Vladimir Dimitrov, who describes that according to fuzzy set theory, the meaning of words cannot be precisely defined. See Dimitrov, *Introduction to Fuzziology*. Also here: http://www.zulenet.com/vladimirdimitrov/pages/fuzzycomplex.html.

557 The six-pointed star is commonly thought of as the Star of David but it was only adopted by Jewish mysticism in the twelfth century due to its association with sex and eastern Goddess religious practice. See Walker, *The Woman's Encyclopedia of Myths and Secrets*, 400-403. I have named it the Star of Aphrodite, as 666 was Her number, though it could equally be known as the Star of Ishtar for similar reasons. See Livingstone, *PaGaian Cosmology*, Appendix D: https://pagaian.org/book/appendix-d/.

many cultural contexts to have been taught that it is foreign. It is not. The autopoietic space in us recognizes Her, this native *Place*, may be comforted by Her, desires Her self-transcendence and self-dissolution. Lammas/Late Summer is an opportunity to be with our organism's love of Larger Self – this *Native Place* – to dwell in this sacred place a while, "make ourselves sacred/whole," which is the meaning of the word "sacrifice."[558] It is not a self-abnegation, it is a fulfillment of purpose, a fulfillment of the passion that is in a being – like the fruit fulfills its purpose in the eating. One's passion, one's work, is fulfilled in the consumption/receiving.

In Pagan tradition this Seasonal Moment has generally been a joyous event, as the focus is on the receiving of the first harvest. It has most frequently been named as "Lughnasad": since in more recent millennia, it has been storied as the wake of Lugh, the Sun King, or Grain God, and it is the Crone that reaps him. He is understood as representing the harvest, thus the burning Straw Man traditions and association with Guy Fawkes. Lughnasad is most often celebrated as an exoteric festival where we do the eating, which is great and holy too, but the esoteric understanding may be a recognition of *ourselves* as the Food; that *we* are the sacred harvest.

"Lammas" is an old name for this Seasonal Moment: this is a Saxon word for the Feast of Bread, the festival of the Great Goddess of the grain, of sustenance.[559] There is rejoicing in this harvest of food, for the sustenance of life. In the pre-Celtic stories of the Goddess tradition, *She* embodied all the transformations – She was the Reaper and the Reaped. She went through all the changes Herself. Within earlier Goddess traditions, all the transformations were Hers;[560] they were transformations of Her aspects. And so,

> … the community reflected on the reality that the Mother aspect of the Goddess, having come to fruition, from Lammas on would

---

[558] See Baring and Cashford, *The Myth of the Goddess*, 161, and also Swimme and Berry, *The Universe Story*, 59.
[559] Walker, *The Woman's Encyclopedia of Myths and Secrets*, 527.
[560] Gray, *The Woman's Book of Runes*, 18.

> enter the Earth and slowly become transformed into the Old Woman-Hecate-Cailleach aspect …[561]

In PaGaian tradition we understand ourselves in this way, as embodying all the transformations, because we do actually. *We* are the harvest – it is our "wake" also. We are the transformations of the ages: that is, we each hold the memory of all Gaian – Earth and Universe – transitions, are indeed the result of those transitions. And we are each the grain that is harvested, we – our lives – are the sacred grain. We are the Reaper in that we each reap our small harvests, and we are the Reaped in that our small harvests are reaped by the Universe – by Her. This is a continuation of the Summer Solstice theme, where we affirmed that we are the Bread of Life. Just as the Food harvested nourishes us, so our lives may nourish the world.

Another indication of the earlier tradition beneath "Lughnasad" is the other name for it in Ireland of "Tailltean Games." Taillte was said to be Lugh's foster-mother, and it was her death that was being commemorated.[562] Tira Brandon-Evans discusses this earlier Goddess tradition in an article about Tailtiu in 2010.[563] There has been suggestion from Goddess scholar Cheryl Straffon that this Seasonal Moment be renamed after Taillte or "Tailtiu": she has suggested *Tailtiunasad*.[564] Many (including myself) have preferred the name of Lammas (instead of "Lughnasad"), but some think it is a Christian term. However, I have understood that Lammas means "feast of the bread," and surely such a feast pre-dates Christianity: She, the Mother of all Life is the original Bread of Life, the sacred Grain and Food of many varieties globally. It is my opinion that Christians simply used the term "Lammas" since it seemed preferable to the clearly Pagan reference of "Lughnasad": that is,

---

561 Durdin-Robertson, *The Year of the Goddess*, 143, quoting McLean, *The Four Fire Festivals*, 20-22.

562 Nichols, "The First Harvest," 1.

563 Tira Brandon-Evans, "Tailtiu: Harvest Goddess," *Goddess Alive!*, Issue 18, Autumn/Winter 2010. http://www.goddessalive.co.uk/issue-18-home/tailtiu-harvest-goddess/.

564 In communications on the Goddess Scholars list, 2010.

use of the term "Lammas" was an appropriation of Indigenous understandings.

The Old One, the Dark and Shining One, has been much maligned, so to celebrate Her can be more of a challenge in our present cultural context. Lammas may be an opportunity to re-aquaint ourselves with the Crone in her purity, to fall in love with Her again, to celebrate *She Who creates the Space to Be.*

Lammas is a welcoming of the Dark in all its complexity: and as with any funerary moment, there is celebration of the life lived (enjoyment of the harvest) – a "wake," and there is grieving for the loss. One may fear it, which is good reason to make ceremony, to go deeper, to commit to the Mother, who is the Deep; to "make sacred" this emotion, as much as one may celebrate the hope and wonder of Spring, its opposite. If Imbolc/Early Spring is a nurturing of new young life, Lammas may be a nurturing/midwifing of death or dying to small self, the assent to larger self, an expansion or dissipation – further to the radiance of Summer Solstice. Whereas Imbolc is a Bridal commitment to being and form, where we are the *Promise of Life*; Lammas may be felt as a commitment marriage to the Dark within, as we accept the *Harvest* of that Promise, the cutting of it. We remember that the Promise is returned to Source. "The forces which began to rise out of the Earth at the festival of Bride now return at Lammas."[565]

Creativity is called forth when an end (or impasse) is reached: we can no longer rely on our small self to carry it off. We may call Her forth, this Creative Wise Dark One – of the Ages, when our ways no longer work.

We are not individuals, though we often think we are. We *are* Larger Self, subjects within *the* Subject.[566] *And* this is a joyful thing. We do experience ourselves as individuals and we celebrate that creativity at Imbolc. Lammas is the time for celebrating the *fact* that we *are* part of, in the context of, a Larger Organism, and expanding into that. Death will teach us that, but we don't have to wait – it is happening around us all the time, we are constantly immersed in the process, and everyday

---

565 McLean, *The Four Fire Festivals*, 22.

566 As Thomas Berry has described the situation of being, in talks he has given.

creativity is sourced in this subjectivity.

As it is said, She is "that which is attained at the end of Desire:"[567] the same Desire we celebrated at Beltaine, has peaked at Summer and is now dissolving form, returning to Source to nourish the Plenum, the manifesting – as all form does. This Seasonal Moment of Lammas/Late Summer celebrates the beginning of dismantling, de-structuring. Gaia-Universe has done a lot of this de-structuring – it is in Her nature to return all to the "Sentient Soup" … nothing is wasted. We recall the Dark Sentience, the "All-Nourishing Abyss"[568] at the base of being, as we enter this dark part of the cycle of the year. This Dark/Deep at the base of being, to whom we are returned, may be understood as the *Sentience* within all – within the entire Universe. The dictionary definition of sentience is: "intelligence," "feeling," "the readiness to receive sensation, idea or image; unstructured available consciousness," "a state of elementary or undifferentiated consciousness."[569]

The Old Wise One is the aspect of the Cosmic Triplicity/Triple Goddess that returns us to this sentience, the Great Subject out of whom we arise. We are subjects within the Great Subject – the sentient Universe; we are not a collection of objects, as Thomas Berry has said.[570] This sentience within, this "readiness-to-receive," is a dark space, as all places of ending and beginning are. Mystics of all religious traditions have understood the quintessential darkness of the Divinity, known often as the Abyss. Goddesses such as Nammu and Tiamat, Aditi and Kali, are the anthropomorphic forms of this Abyss/Sea of Darkness that existed before creation. She is really the Matrix of the Universe. This sentience is ever present and dynamic. It could be understood as the dark matter that is now recognized to form most of the Universe. This may be recognized as Her "Cauldron of Creativity" and celebrated at this Lammas Moment. Her Cauldron of Creativity is the constant flux of all form in the Universe – all matter is constantly transforming. *We* are

---

567 Doreen Valiente, "The Charge of the Goddess" cited in Starhawk, *The Spiral Dance*, 103.

568 Swimme, *The Hidden Heart of the Cosmos*, Chapter 13.

569 *Webster's Third International Dictionary of the English Language.*

570 Berry, *Evening Thoughts,* 149.

constantly transforming on every level.

These times that we find ourselves in have been storied as the Age of Kali, the Age of Caillaech – the Age of the Crone. There is much that is being turned over, much that will be dismantled. We are in the midst of the revealing of compost, and transformation – social, cultural, and geophysical. Kali is not a pretty one – but we trust She is transformer, and creative in the long term. She has a good track record. Our main problem is that we tend to take it personally.

The Crone – the Old Phase of the cycle, *creates the Space to Be.* Lammas is the particular celebration of the beauty of this awesome One. She is symbolized and expressed in the image of the waning moon, which is filling with darkness. She is the nurturant darkness that may fill your being, comfort the sentience in you, that will eventually allow new constellations to gestate in you, renew you. So, the focus in ceremony may be to contemplate opening to Her, noticing our fears and our hopes involved in that. She is the Great Receiver – receives all, and as such She is the Great Compassionate One. Her Darkness may be understood as a Depth of Love. And She is Compassionate because of Her dismantling … where we may not have the will. We may want to be ever fresh, ever new, yet it is not possible without the Wise Old One, who will mercifully shake us loose from our tracks. Often we have only feared Her. Sometimes the changes that need to be made are awesome … we would not have chosen them, but they serve us deeply. The Zen Buddhist tradition speaks of the "tiger's kindness," that is, we *want* to change, but may not have the will. (The tiger fears the human heart, the human fears the tiger's kindness).[571]

This Seasonal Moment is about trusting and rejoicing in the kindness and Creativity of the Dark – knowing it is centrally part of us and we are part of it. Loren Eiseley describes his pulse as "a minute pulse like the eternal pulse that lifts Himalayas and which, in the following systole, will carry them away."[572] Our organisms are constantly a microcosm of the cataclysmic transformations of Gaia – transformations that allow the life of the organism to go on, be that our small self or that

---

[571] Susan Murphy, *upside-down zen: a direct path into reality* (Melbourne: Lothian, 2004), 89-93.

[572] Eiseley, *The Immense Journey*, 20.

of the Large Self of Earth or the Universe.

Lammas/Late Summer Breath Meditation

To begin this Seasonal Moment's contemplation:

> Centering for a moment … feeling the full cycle of your breath. Recognising the Triple Goddess Dynamic here in this primary Place: She who unfolds the Cosmos … receiving the Gift, as you inhale – feeling the Urge to Be, feeling the peaking – the exchange, the communion that this Place is, then the release – the dissolving, becoming the Gift, making space, ever transforming: the Triple Spiral dynamic here in your breath, the Wheel of the Year in your breath.

Some Lammas/Late Summer motifs

[Figure 26]

Suggested decorations for this Seasonal Moment – on the front door, windowsills, corners of shelves, and/or altars, may include:

- **grain and produce from garden** – representing the harvest, food.
- **ash** – representing the dark sentience whence we come and to whom we return.
- **cauldron** – representing the *Cosmic Pot* in which we are immersed, the Larger Self, the Recipe/Feast of which we are part.
- **bread figures** – representing the individual being, the self, who is consumed; it is the *Sacred Consuming.*
- **black of different textures** – representing the beauty of the dark.
- **new seed pods** – representing the new dark.
- **autumn/ripe produce coloured yarn** – representing the harvest.

An Invocation

Begin with the ringing of a bell, if possible: it is in contrast to the ringing of the small bell in Imbolc ceremony, when it represents the advent and unique beauty of the small self. The Lammas bell should be larger and recalls the passing of the small self into the Larger Self, which is being celebrated in this beginning of the dark part of the cycle. Note that this passing happens in many ways throughout our lives, not just at death.

After ringing the bell perhaps three times, pronounce:
for whom does the bell toll? Then answer: *It tolls for thee.*
Continue:
Breathe this breath that is yours – as if it were your first.
Breathe this breath that is yours as if it were your last.
Breathe this breath that is yours, and not yours.

This is the Truth – beyond all knowledge, which the Old One represents.

The Universe Story and Lammas

The story of the universe may be described as "a story of majesty and beauty as well as of violence and disruption, a drama filled with both elegance and ruin."[573]

The process of evolution may be named the "Transformation-of the-Ages": it is a name I give to the Old One/Crone. Transformation may be understood as a power of the universe, as cosmologist Brian Swimme speaks of it in his DVD series *The Powers of the Universe.*[574]

Another term that I have used to describe the poiesis of this Season of Lammas is "All-Nourishing Abyss." It is Brian Swimme's term for the mystery at the base of being,[575] and it is a term I use to speak of the Dark Source to whom the Old One returns us (can They be distinguished?)[576] The All-Nourishing Abyss is both generative and infinitely absorbing, a power out of which particles spontaneously emerge and into which they are absorbed, not just thirteen point seven billion years ago, but in every moment.[577] We may imagine this Power as the Great Receiver, the Old Compassionate One, complete forgiveness, Transformer, Depth of Love. Swimme says that "the foundational reality of the Universe is this unseen ocean of potentiality."[578] He speaks of it as "nonvisible" (as distinct from "invisible"): that is, as "that which can never be seen."[579] I have used the term "manifesting" reality to speak of this dark, from which the "manifest" emerges, and into which it is absorbed.[580] Swimme says that "the appropriation of the new cosmology depends upon an understanding of the reality and power of the nonvisible and nonmaterial realm."[581] As Swimme describes: "All-nourishing abyss then is not a thing, nor a collection of things, nor even, strictly speaking, a physical place, but rather a power that gives birth and

---

573 Swimme and Berry, *The Universe Story*, 47.

574 Program 8, "Transformation."

575 Swimme, *The Hidden Heart of the Cosmos*, 100.

576 I contemplate and discuss this question in "Contemplating Triple Goddess as Creative Triad, a Depth of Love": https://www.magoism.net/2017/08/essay-contemplating-triple-goddess-as-creative-triad-a-depth-of-love-by-glenys-livingstone-ph-d/.

577 Swimme, *The Hidden Heart of the Cosmos*, 100.

578 Ibid.

579 Ibid., 97.

580 As noted in the Introduction to this book, along with references.

581 Swimme, *The Hidden Heart of the Cosmos*, 97.

that absorbs existence at a thing's annihilation."[582] So, is my understanding of the Womb of Space we inhabit, and to whom we return.

Lammas celebrates the beginning of dismantling, de-structuring, cutting the harvest, after the peaking and ripening of Summer Solstice. There are many such moments in the evolutionary story which could be specifically remembered, including our present ending of the Cenozoic Era, as Swimme and Berry describe the present extinctions and planetary destruction.[583] Gaia Herself has done a lot of this de-structuring, it is in Her nature to return all to the sentient "Soup," for creative purpose, for never-ending renewal.

And a quote from *The Universe Story*, that is resonant with Lammas/Late Summer:

> Eventually, in a million years or in several billion years, a star's resources against the collapse are all used up. If the mass of a star at this point is large enough, its gravitational pressures will destroy the star. The remaining materials will rush toward each other. Nothing in the universe can now stop them. ... This stellar being that burned brightly for billions of years, that may have showered sentient creatures with radiant energy that they transformed into their living bodies and into cathedrals that rose in wheat fields, has gone, only a black cinder left.[584]

Notes on Lammas/Late Summer Metaphor

The cauldron may be featured and understood to represent the cosmic sentient soup in which we are immersed. We remember to whom the harvest of our lives belongs: that is, to the Cosmos, the larger Creative Cauldron, the real "World Bank" from which all draw and return.[585] The

---

582 Ibid., 100.

583 Swimme and Berry, *The Universe Story*, 241-250.

584 Ibid., 48.

585 Helpful suggested texts to refer to are: "Cauldron" in Walker, *The Woman's Encyclopedia of Myths and Secrets*, 150-154, "Cerridwen" in Stone *Ancient Mirrors of Womanhood*, 58-59, "Cerridwen: Queen of Wisdom" in Edwards, *The Storyteller's Goddess,* 151. (NOTE: Cerridwen is often

Lammas cauldron is for dis-solution, the Samhain "Womb" is for re-solution, though the two are not really distinguishable: there is simply different qualities of the same creative Place.

Re "making sacred"/sacrifice: My understanding is best explained with this excerpt from my personal journal and doctoral thesis:

> It seems to me that this "making sacred" of the Goddess, is very different from the "sacrifice" demanded by the Christian cosmology – that is certainly my experience. The kind of "sacrifice" demanded by the father god of Christian cosmology, left nothing for me – it was an abnegation of myself. Whereas this current experience of making my life sacred, is an acknowledgement of the Source of my Creativity, the One to whom we belong. So it is not taking something from me, it is remembering the roots of me – where I am planted. And my fullness is Her Harvest – the more the merrier. My desires are Her desires ... they are one. This is very different from having to submit to some tyrant, and forego what is in me!

This is about "restoring wholeness" – small self becomes whole in Larger Self. Goddess cosmology does not need blood sacrifice[586] … Hers is a "menstrual cosmology."[587] Can we begin to think about these things from within our own skins: that is, with "maternal minds," or with minds as if we (especially females) mattered?

Whereas at Imbolc, we shone forth as individual, multiforms of Her; at Lammas, we small individual selves remember that we are She (in case we had forgotten), and dissolve back into Her. We are the Promise of Life as we affirmed at Imbolc, but we are the Promise of *Her* – it is not ours to hold. We become the Harvest at Lammas; our individual harvest *is* Her Harvest. We are the process itself – we are Gaia's Process.

---

associated with Samhain as "conceiving space, re-solution, but may also be associated with Lammas as creative space for dissolution.)

586 As discussed in Chapter 4 Samhain/Deep Autumn. See Baring and Cashford, *The Myth of the Goddess*, 160-161 where blood sacrifice is discussed.

587 Raphael, *Thealogy and Embodiment,* 270.

It is a surrender – to Her, which is much safer than as surrender is often experienced.

Re understanding Her as the Compassionate One, deeply committed to transformation: She is merciful, because frequently we would not have the stomach for what it may take to create the Space to Be, and also She receives all, thus *forgives* all.

This process of ceremonial celebration of the whole wheel of the year will include and invoke a revaluing of the dark, especially in this Seasonal Moment of Lammas. Lammas is a particularly powerful Moment on the wheel, partly because of the cultural alienation from this essential aspect of life/cosmic creativity. Thus also, care needs to be taken in the Lammas process, as one may have a great deal of negative stories about "darkness" which will affect experience.[588]

Notes on imagery and metaphors in the offered Ceremonial Script

In the Centering and Breath Meditation, the ringing of the bell is meant to recall funeral rites, and the focus is on the passing of one's individual breath, the breath "that is yours, and not yours," how we pick the breath up at birth, let it go at death, we/all borrow it for our particular lifetime. We breathe a collective breath, that belongs to All. The air remains, altered by your (our) presence and actions, for those who come next.

Re Calling the Directions/Casting the Circle: the focus is on identifying with the elements as part of the Old One's "Recipe." Small clay pots in cauldron shape for each of the elements may be helpful. Each participant affirms their elemental presence in "Her Cauldron of Creativity": that is, the all-pervading constant flux of all matter in which we are immersed[589] – the Soup of the Universe. The suggested solemn paced drumbeats for a few moments may echo the tempo and mood of the passing of all.

---

[588] Some notes on my personal process are available in *Evolution of the Ritual Scripts,* Chapter 7 of my Ph.D. thesis, *The Female Metaphor – Virgin, Mother, Crone – of the Dynamic Cosmological Unfolding: Her Embodiment in Seasonal Ritual as Catalyst for Personal and Cultural Change.*

[589] This constant flux of all matter in which we are immersed has been described scientifically by Vernadsky, *The Biosphere*, 34.

Regarding the Invocation: whereas in Imbolc ceremony we affirmed that we are the shining and New One – ever the Urge to Be – the beauty of the particular and emerging self, now at Lammas we may affirm that we are the Dark and Ancient Wise One – ever in the process of dissolving and recycling, creating the Space to Be, and indeed formed by all that went before. We are as ancient as Gaia Herself, ever She who has transformed over and over. Ten percent of our bodyminds are hydrogen – direct from the primordial fireball, the rest made in our grandmother supernova (Tiamat as She may be named[590]) – recycled over and over. We hold all of the evolutionary wisdom within our cells – Her ancient story is your/our story. Participants receive grains of barley/wheat/native grain in their hands and affirm that they are the grain that is harvested. They receive ash on their foreheads, to remember Her dark sentience within, and affirm that they are: "She, *Dark and Ancient Wise One*"; adding or expressing their own words to this theme. A short meditative space is included for contemplating Her, this dark and ancient sentience within, for which participants may choose to wear masks or cover their heads with a dark veil, for possible enhancement of the sense of identification with Dark Larger Self.

Note here, that it is not common for the depths of these particular invocations to be understood in most present cultural contexts, and it is does make a difference if the one speaking the invocation to another knows some depth of what they are saying: so the decision about whether to pass the blessing around the circle or have it spoken by one or two celebrants to all, should depend on the experience of the group.

Each participant makes and brings a bread figure,[591] which will be given to the fire in the ceremony (or a large bowl of water), with an offered dedication. The One to whom the bread figure is given is understood as the Compassionate One, She who receives all, and transforms all. So much of our being yearns for this compassion and "forgiveness" (which is a "receiving"), and often yearns too for transformation. We have to be able to trust that there is a much bigger picture that we don't usually see. So, we trust Her knowing, which is

---

590 See Chapter 5 Winter Solstice/Yule for that story.

591 See Appendix I for a salt dough recipe. Also here: http://www.ancientnile.co.uk/saltdough.php.

beyond all knowledge. Each may perhaps contemplate what holds them back from letting go to deeper self and transformation, what are the fears. This rite is meant as a return/dedication to a Beloved One, a "wholing," to nurture healing relationship with the Dark in whom we are immersed and sustained by.

In the script as it is offered here, the group chants[592] and repeats a summarised form of each one's expressed dedication: this is to reflect them back to the one speaking, and affirm the hearing by all, and the self. In this process each one may discover other dimensions to their considered responses.

The (five pointed) star carrots that are offered to each participant as representations of "hoped for harvests" may be star cookies instead. I choose carrots because everyone can usually eat them, and the orange colour is cheerful, but gluten-free cookies may be enjoyable. This "Hoped for Harvest" rite/process is based on one in Starhawk's *The Spiral Dance.*[593]

For Communion at Lammas, dark bread and dark beer/wine /juice may be offered to participants, with blessings for being nourished by Her Harvest, and to *be* nourishing. The dark drink may be understood as a reference to our situation in the Womb of the Mother wherein Her darkness and blood is for gestating, is for life.

There is a circle dance that may be done after the Story sharing, before Opening the Circle. I name it the "Harvest Dance"; details are noted in the offered script. For contemplation as we dance, questions are asked before: "Can the grub imagine the butterfly she will become? Can we imagine what will emerge?"

---

592 The chanting is based on the process described in Starhawk, *The Spiral Dance*, 207.

593 Starhawk, *The Spiral Dance*, 208.

[Figure 27] A Lammas/Late Summer Ceremonial Altar

A LAMMAS/LATE SUMMER CEREMONIAL SCRIPT

Note that the directions here are called in "counter-clockwise" direction which is sunwise in the Southern Hemisphere. The order may be changed to "clockwise"; sunwise for the Northern Hemisphere. Note that this particular offered ceremonial script is not considered prescriptive, and that there are many creative possible variations.

Some requirements and some suggestions: Each bring a bread figure that they made with contemplation on what they would have transformed – perhaps fears, what they would "make sacred," what they have for-giving: keep bread figures with them – take to pre-ceremonial contemplation. Optional mask/dark veil for Contemplation of Dark Larger Self. Wear black preferably, or dark colours, and bring dark veil and/or wear a dried garland. Black altar cloth, perhaps with varied textures that accentuate the beauty of black. Dark rye bread, perhaps drizzled with olive oil. Dark beer, red wine and/or dark grape juice. Dark foods such as licorice, prunes, black olives. Waterbowl or fire ready. Empty basket lined with black cloth for bread figures. Star carrots/cookies in a small basket nearby. Small bowl of ash. Small basket of grain. Cauldron-like clay pots for all the elements – brass pot of methylated spirits within it for fire. Perhaps add cloves/dark spice to water element. Wreath around centre candle: dried roses/flowers still on it from Summer is good, with added wheat/grain heads, rosehips, and fresh seed pods. Music ready.

Call to Gather – drumming with a slow solemn beat or a medium size bell, slowly ringing.

Centering and Breath Meditation
"For whom does the bell toll?" asked three times, with a different emphasis on *toll*, *bell* and *whom* each time; then the answer stated once, "It tolls for thee."

Breathe this breath that is yours, as if it were your first … do you remember that it was so? Breathe this breath that is yours, as if it were your last ... do you remember that it will be so? Breathe it now, this breath that is yours ... but not yours. We breathe this breath together (take hands) … we are the breath of the Cosmos in this time and in this place – Her breath-taking manifestation, in this Moment of Her story." (let go hands)

Statement of Purpose

This is the season of the waxing dark. The seed of darkness that was born at the Summer Solstice now grows; the dark part of the days grows visibly longer. Earth's tilt is taking us back away from the Sun. This is the time when we celebrate dissolution, the time when each unique beautiful self lets go, to the Darkness. It is the time for celebrating ending, when the grain, the fruit, is harvested. We meet to remember the Dark Sentience, the All-Nourishing Abyss, She – from whom we arise, in whom we are immersed and to whom, we return.

This is the season of the Crone, the Wise Dark One, who accepts and receives our harvest, who grinds the grain, who dismantles what has gone before. She is Hecate, Lilith, Medusa, Kali, Ereshkigal, Chamunda, Coatlique – Divine Compassionate One, She Who Creates the Space to Be. We meet to accept Her transformative embrace, trusting Her knowing, which is beyond all knowledge.

Let us slip between the worlds into Her dark space. Let us begin by remembering Her Recipe, the elements in Her Cauldron from whence we come.

Calling the Directions/Casting the Circle
Celebrant: "Hail the East: Water we are – filled with the ocean tides – in Her Cauldron of Creativity!"
Drum for a minute to contemplate this
Co-celebrant presents each around the circle with the cauldron of water, ladles it into their hand; each may taste it, hold their wet hand to the East and respond, affirming "Water I am! in Her Cauldron of Creativity" (and whatever else they may like to add).

Celebrant: "Hail the North: Fire we are – sparks of ancient heat – in Her Cauldron of Creativity!"
Drum for a minute to contemplate this
Co-celebrant lights the firepot (methylated spirits in brass pot, or sand in clay pot with methylated spirits poured on – both are soft flames. (A lid to put over the brass pot when done – if that is the chosen method – is good: that will extinguish the flames.)
Each participant comes to the firepot, lights a match, holds it to the North, and responds, affirming "Fire I am! in Her Cauldron of Creativity" (and whatever else they may like to add).

Celebrant: "Hail the West: Earth we are – geoformations – in Her Cauldron of Creativity!"
Drum for a minute to contemplate this
Co-celebrant presents each around the circle with the cauldron of earth; each takes some and holds it to the West and responds, affirming "Earth I am! from Her Cauldron of Creativity" (and whatever else they may like

to add).." Each puts the earth back in the pot and may touch their "dirty" hand to own forehead/throat/chest.

Celebrant: "Hail the South: Air we are – this ancient river that all have breathed – in Her Cauldron of Creativity!"
Drum for a minute to contemplate this
Co-celebrant presents each around the circle with the cauldron with a lit smudge stick in it; each lifts the smudge stick out, holds it to the South and responds, affirming "Air I am! in Her Cauldron of Creativity" (and whatever else they may like to add)." Each puts the smudge stick back in the pot.

Celebrant: "This is Her Recipe – Water, Fire, Earth and Air (PACING THE CIRCLE); we are made from this, these are our Origins. We are at the Centre of Creativity, in Her Cauldron: we are immersed in it, in this *Dark Realm of Manifesting*. Her Centre is Here. The circle is cast, we are between the worlds, beyond the bounds of space and time, where light and dark, birth and death, joy and sorrow meet as one.
(light centre candle)

Invocation
Celebrant: "Let us invoke and recognize Her now, in ourselves and in each other – the Ancient and Dark Creative One, this Wise One, this Larger Self whom we are."
Celebrant (with basket of wheat/barley grain) to each, or to one participant: "(name) ***you*** are the grain that is harvested – you are Her Harvest" (puts grain in their hand).
Response: It is so. I am … (each chooses whatever they wish to repeat and affirm for themselves, or may add or use own words)
The basket of grain may be passed around the circle, with each repeating the blessing/invocation.

Celebrant (with small bowl of ash) to each, or to one participant: "Remember Her dark sentience within you (touch their forehead with ash) … that you are the Ancient Wise One, you are the Transformation of the Ages."
Response: It is so, I am … (each chooses whatever they wish to repeat

and affirm for themselves or may add or use own words).
The bowl of ash may be passed around the circle, with each repeating the blessing/invocation.

Contemplation
Let us sit and contemplate Her now within us – this *Dark and Ancient Wise One*, this Larger Self whom we are.

All may put on their masks if they wish or cover their face with their dark veil.

MUSIC[594] or slow drum optional, for a few or several minutes, in which time the celebrant puts on her mask, and/or may add a dark veil to cover her face – as desired.
Celebrant announces: "All stand – put your grains in the centre, take up your bread figures."

Dedication/Expansion – the Bread Figures
Celebrant paces the circle with the empty basket announcing: "I am the Ancient Creative One within you and within all – the Compassionate One who receives all, She who transforms all."

Celebrant ends her pacing at the fireplace/waterbowl, and puts the basket down, announcing: "Now is the time for your dedication to Deep Self. Step forward, one at a time, as you are ready, offer your bread figure, and speak if you wish. What would you surrender for transformation? What do you have for-giving? Perhaps what do you want to 'make sacred' – dedicate to deep wisdom within?"

Each one comes forward when they are ready, speaking if they wish, creating their own words or may choose to speak some of this offered dedication, or a variation of them:
I dedicate my small self to Larger Deeper Self, understanding that I am She – GaiaUniverse – She Who is All. I am the beauty of the Dark: the Sentience of the

---

594 I have used an excerpt from *Ignacio* by Irene Papas and Vangelis Papathanassiou, *Odes* CD (1979), which goes for about 4 minutes.

Cosmos is within me, and I am in Her. I trust Her Wisdom and Creativity – She Who receives me in every moment, so that new composition may arise. I offer Her Darkness … for transformation. I have … for-giving. I want to make sacred …

The participant then places their bread figure in the basket; and then repeats whatever they wish in summary of their dedication, chant style. The group repeats the answer in chant style, perhaps three times, or it may be enhanced with a drum and carried on to raise energy for each.

Those choosing not to speak may have the drumming for their intention.

Celebrant takes her turn also, first removing her mask and/or veil ceremonially (this may be done by the celebrant walking the circle in opposite direction from original pace, taking off the mask and veil, and returning to the circle). A co-celebrant then walks the circle and takes up the mask and/veil.

WHEN ALL ARE DONE
Celebrant returns to the basket near the fireplace/waterbowl and announces: "The Old Compassionate One – the Creative One receives all, transforms all. We are Her Harvest in every Moment."

Celebrant tosses the bread figures into water/fire slowly, one at a time, with the blessing: "For transformation and dedication."
The group may join their voices to the celebrant's.

When all the bread figures have been dedicated, the group affirms: "May it be so."

Silence
Celebrant: "Everything passes, all fades away. May we open our Hearts."
(A few moments of silence as all watch the bread figures burn/dissolve.)
Chant: "She lies beneath all. She covers all."[595]

[595] Based on a chant in Starhawk, *The Spiral Dance*, 195.

"I am Love"

Celebrant (holding the basket of stars) announces:
"She says: I am Love, I am the All-Nourishing Abyss. All manifestation springs from me – you have given yourself back. You will reap the harvest, you will proceed in joy and abundance … for this is the Mystery."

All respond: It is so!

The Hoped for Harvest

Celebrant takes the basket of star carrots/cookies around the circle asking each.
"What do you hope to harvest? What do you hope for? What will nourish you?"

Each person answers as they wish, then repeats whatever they wish in summary, chant style. The group repeats the answer in chant style, perhaps three times, or it may be enhanced with a drum and carried on to raise energy for each.

Those choosing not to speak may have the drumming for their intention, after they take a star.
A co-celebrant asks question for the celebrant.
WHEN ALL ARE DONE
Celebrant holds up their star announcing: "May all these hoped for harvests be so."
All respond: "May it be so!" All eat their stars.

Communion

Celebrant holds up the plate of dark rye bread and the decanter of dark beer, announcing: "We are filled with and nourished by the *Creative Dark* – immersed in Her Womb. May we receive Her nourishment and be nourishing."
Celebrant offers the bread to a person in sunwise direction, with the blessing: "May you receive Her nourishment and be nourishing."

The plate of bread is passed around the circle, with each blessing the next, in sunwise direction.

Someone distribute glasses to each person.
Celebrant passes the dark beer/dark juice sunwise to one in the circle with the blessing: "You are filled with and nourished by the Creative Dark – immersed in Her Womb. May you receive Her nourishment and be nourishing."
Each repeats the blessing around the circle, as they offer the beer/wine/juice.

Other dark foods may be offered, as all sit and relax.

Stories (with ceremonial manners)
Let us tell each other stories – perhaps of things you are letting go of, perhaps stories of loss, or expansion into deep self; perhaps stories of harvests, transformations hoped for. Or whatever you would like to say at this time, and we will listen.
The group listens to each – it is not a conversation or a space for comments: and all give a formal response to each.

Dance – Harvest Dance[596]

---

[596] The dance that may be done is based on one that I learnt from Jean Houston, and I name it the "Harvest Dance": form a circle holding hands, take three steps to the right, pause, then three more steps to the right, pause and sway LRL. Take three more steps to the right, pause, take two steps into centre with right foot each step, two steps back out with left foot each step, the sway L. Then begin again with the steps to the right. Add gathering motions with joined hands as you make the steps into the centre.
Music: "Menousis," *Odes* CD, Papas and Vangelis. More about the dance here: https://pagaian.org/2022/01/30/harvest-dance-for-lammas-ceremony. An alternative could be an improvised dance to "Dance me to the End of Love" by Leonard Cohen or The Civil Wars version: https://thegirlgod.com/pagaianresources.php.

"Let us dance in celebration of the Harvest – our harvests and the Harvest of these times. Can the grub imagine the butterfly she will become? Can we imagine what will emerge?"
PAUSE AT FINISH of the dance, holding the circle for a few moments.

Open the Circle
Celebrant: "All turning to South:
We have remembered this evening that we are Her Recipe –
that we are Air, in Her Cauldron of Creativity.
May there be peace within us."
All respond: May there be peace within us.

Celebrant: "All turning to West:
We have remembered this evening that we are Earth in her Cauldron of Creativity.
May there be peace within us."
All: May there be peace within us.

Celebrant: "All turning to North:
We have remembered this evening that we are Fire, in Her Cauldron of Creativity.
May there be peace within us."
All: May there be peace within us.

Celebrant: "All turning to East:
We have remembered this evening that we are Water, in Her Cauldron of Creativity.
May there be peace within us."
All: May there be peace within us.

Celebrant: "All turning to Centre:
We have remembered this evening that we are Sacred and Ancient Wise One, that we are the Grain cut for Harvest … that we are the Transformation of the Ages. We have remembered the Dark Sentience, the All-Nourishing Abyss, this Larger Self to whom we belong, in whom we are immersed, and to whom we return. We have dedicated ourselves

to Deep Wisdom within, and expressed our hopes. We have remembered that our harvest is Her Harvest.
May there be peace within us and between us." (all join hands)
All: May there be peace within us and between us.

Celebrant: "The circle is open but unbroken. May the Wisdom and Compassion of Goddess – the Mystery – go in our hearts and minds. It has been a merry meeting. It is a merry parting. May we merry meet again. Blessed Be!"

Music and Dance option
Suggest: "Dance me to the End of Love" and "Lady Midnight" by Leonard Cohen.

Points for individual contemplation prior to the ritual:
- How you are the grain that is harvested, how you are Her harvest.
- The dark sentience that fills you, in which we are immersed. How you are the Transformations of the Ages, the Ancient One.
- Your bread figure that you give to the Old Dark One's basket and what you might like to express – desired transformations, fears perhaps, dedication to deep wisdom, Larger Self beyond your knowing. You might understand it as a "for-giving," or a "making sacred" – whatever feels right for you at this time.
- What do you hope to "harvest" – what do you hope for?

Lammas/Late Summer Goddess Slideshow:
https://thegirlgod.com/pagaianresources.php

CHAPTER 11
AUTUMN EQUINOX/MABON

Southern Hemisphere – March 21-23,
Northern Hemisphere – September 21-23

Autumn Equinox/Mabon is the midpoint between Summer and Winter Solstices in the dark part of the annual cycle: it is the balance point in that part of the cycle. The young and growing dark has come into balance with the light. The Sun is equidistant between South and North on the horizon, yet the trend at this Equinox is toward increasing dark and cold: thus it may be understood as a "stepping into the power of loss," as manifest life continues its decline at this time. Autumn Equinox is a time of thanksgiving for the harvest of the passed year/years – for its empowerment and nourishment, and it is also a time of leave-taking and sorrow, recognition of the cost of the harvest, the losses involved, as life declines. It is a moment of certain descent, feeling the loss and grief, which may be a journey to wisdom, vision and sovereignty.

In the PaGaian wheel Autumn Equinox is understood as a point of balance of the three qualities of Goddess, of Cosmogenesis: the delicate balance of these qualities that keeps it all spinning. It is not a simple duality. All at once, the three faces of Cosmogenesis are present. As Seed, She is both sovereign of the underworld – *Old Wise One* fertile with knowledge (the autopoiesis of the Cosmos), and at the same time the irrepressible *Urge to Be* (unique differentiated being); and She is the *Mother, Source of Life* (the communion of the Cosmos). As Goddess, She may be named Persephone/Kore, Demeter/Ceres and Hecate – all three in each other in a complex triplicity. This is a blessed Moment of a harmony/balance that streams through the grief and the ecstasy of life. The Seed may represent all *Three* as they are inseparable, a holy trinity, and the sacred balance.

The Equinoxes are mid points of pause between Solstices, and they are traditionally the points of celebration of the Eleusinian Mysteries – the 'Lesser' in the Spring, and the 'Greater' in the Autumn. *The Mysteries of Eleusis* was a nine-day ceremonial celebration of Mother and Daughter, named as Demeter and Persephone, that took place annually for almost two thousand years, at Eleusis in Greece,[597] and they grew out of the much older autumn festival, the *Thesmophoria*, a three day festival timed exactly with the dying and resurrection phase of the lunar story.[598] The rites of the *Thesmophoria* were "dedicated to Demeter at the time of the sowing of the seed."[599] The Eleusinian Mysteries were thought to hold the entire human race together – partly because people came "from every corner of the earth to be initiated."[600]

Some More Autumn Equinox Story

In PaGaian Cosmology, the story told of Persephone's descent is based on Charlene Spretnak's version,[601] wherein Persephone descends of Her own volition; She is not forced/taken. She descends with the desire to "tend the dead." In this pre-patriarchal story Persephone may be understood as a 'hera,' a pre-Hellenic name for Goddess which pre-dates the masculine form 'hero,' according to Charlene Spretnak's analysis.[602] Persephone's descent to the Underworld for the tending of the dead may be understood as the tending of *grief and sorrows*, which She is able to do because as *Seed*, a numinous reality, She represents rejuvenation, the restoration of beauty and communion. This would have been the integrity with which this Seasonal Moment was apparently initially celebrated. In Carolyn McVickar Edwards' version of the story,

---

597 In the fourth century C.E. they "were proscribed by the Christian Theophrastus, and later the temples were sacked by the Goths." Baring and Cashford, *The Myth of the Goddess*, 374.

598 Baring and Cashford, *The Myth of the Goddess*, 385.

599 Baring and Cashford, *The Myth of the Goddess*, 374. The festival celebrated Demeter as law-giver; She was "the goddess who gave the law that transformed Greece from a nomadic into an agricultural community," Ibid.

600 Durdin-Robertson, *The Year of the Goddess*, 158.

601 Spretnak, *Lost Goddesses of Early Greece*, 111-117.

602 Spretnak, *The Politics of Women's Spirituality*, 87. Charlene comments that 'hero' is a term for the brave male Heracles, who carries out the bidding of his Goddess Hera, and that the derivative form 'heroine' is completely unnecessary. She suggests using 'hera" for everyone who acts courageously, and so I do.

Persephone descends for the gaining of wisdom.[603] For millennia, in Greece, it was the holy celebration of Persephone's descent to the Underworld, and in the earliest Goddess tradition, in pre and post-patriarchal understandings, Her *voluntary* descent corrects the narrative, returns Her agency; that is, She simply understands the necessity of the journey, a journey to self-knowledge and redemption of the lost. As such She is shaman, redeemer.

In pre and post-patriarchal understandings, Persephone enacts the integrity and status of *Wisdom Redeemer* figure within all being. She, like the seed, is the Mother knowledge who grows within.[604] This relationship of Child/Daughter with the Mother-self has traditionally (in Pagan circles) been represented in the apple, when it is cut across the diameter, revealing seed in the fruit in pentacle shape: and it is said that She is "seed in the fruit, that becomes fruit in the seed."[605] Persephone/Kore (Daughter/Child) is within the heart of the apple (Mother). This is a religious relationship: the essence of the Mother-Daughter Mysteries of Old, and wherein all initiates (women, men and children) identify with the Mother and also the Daughter.[606] Each being is indeed both: born and made of Her *materia.*

The beauty and sustenance which we may celebrate and enjoy daily unfolds from the Underworld, the deep dark earth – into which the seed goes. Autumn Equinox/Mabon Seasonal poetry may express a stepping into the *Power of the Abyss,* which is a descent to wisdom.

---

603 Edwards, *The Storyteller's Goddess*, 178-183.

604 See *The Equinoxes as Story of Redemption: Sacred Balance of Maternal Creativity* by Glenys Livingstone, http://pagaian.org/2015/03/17/the-equinoxes-as-story-of-redemption-sacred-balance-of-maternal-creativity/, based on *Female Metaphor, Science and Paganism: a Cosmic Eco-Trinity* in The Indian Journal of Ecocriticism, available at https://pagaian.org/articles/female-metaphor-science-and-paganism-a-cosmic-eco-trinity/.

605 As is said of each participant in the offered Autumn Equinox ceremonial script in this chapter, based on Starhawk, *The Spiral Dance*, 212 (for Samhain ceremony).

606 See Rachel Pollack, *The Body of the Goddess* (Brisbane: Element Books, 1997), 220-221.

Lawrence Durdin-Robertson describes the beatific significance of the showing of the "ear of corn" (as the wheat was known), and the grain's vulval shape. He calls it a 'Vision into the Abyss of the Seed,' a vision of the Source of Life.[607] The vision was/is at the same time a vision of the Daughter/Persephone and continuity, the *fact* of renewal (not just "hope"). The vision was an initiation into knowledge of the seed, as a thread of life that continues beneath the visible. The grain of wheat (or rice or corn in other cultures) may represent what we are given in every moment, daily sustenance. To be aware of this gift, the gift of the present moment in its depth, is sacred knowledge – Divine Wisdom.

In this cosmology as it has been expressed, where Lammas/Late Summer was a celebration of the Abyss itself in, an assent to the process and the beauty of dissolution, this Seasonal Moment of Autumn Equinox/Mabon is about *feeling* it – engaging with it, knowing its pain and its gain. It is a harvest moment, so there is deep gratitude; but there is a cost and loss involved, so there is deep grief. The classic mask expressing a face of half joy and half grief is appropriate to this Moment (see Figure 24).

The name "Mabon" that is frequently used for this Seasonal Moment is Celtic (or even pre-Celtic?). It has had, and does have, other names, in different regions and cultures: in Wales it has been known as "Alban Elfed."[608] "Mabon" is the name of the Son of Modron – the Matrona or Mother of earliest times. His name is not a name but is a title "Mab ap Modron" meaning "Son of the Great Mother."[609] Mabon is taken from Her when only three nights old as the story goes in the Mabignogion. "He is the primal child who was in existence at the beginning of things …," as Caitlin and John Matthews note.[610] He represents an innocence; and note here that "innocence" is understood as being a connection to the Mother, to Source. It is a quality of direct connection to Cosmos – not in any arrogant way, but in a "selfless" way: it is a "Don Quixote" thing. In the still strong mysteries of Mabon and

---

607 Durdin-Robertson, *The Year of the Goddess*, 166. Also see Baring and Cashford, "The Ear of Wheat," *The Myth of the Goddess*, 389.
608 See Orr, *Spirits of the Sacred Grove*, 234-235.
609 French, *The Celtic Goddess,* 138-139.
610 *The Western Way*, 83.

Modron the Mother, he is lost and imprisoned, but his primal innocence is held to turn away harm. The reason for this is his continuing connection to the Mother; like Persephone in the Greek story, he is a thread that continues. Though he is lost he has the power of protecting life; he IS life (as Persephone is). One could choose to focus more on this story in Autumn Equinox ceremony.

The oldest story of descent and return that may be known by a significant portion of humanity is that of the Sumerian Goddess Inanna, written down by High Priestess poet Enheduanna around 2300 B.C.E..[611] Inanna was the "primary one" of Sumer for three and a half thousand years and made the heraic journey into the Underworld gaining knowledge of that realm. Her story is based on the descent and return of the Moon, as many other later death and resurrection stories are: the Moon was the original primordial vision of redemptive power, expressed within a three day process[612]. There is also the story of the descent of Dumuzi, Inanna's Beloved – for whom She grieves.

Like its Spring Equinox counterpoint, Autumn Equinox may be expressed as a "stepping into power," as the balance tips, but this time it is a descent, a stepping into the power of loss and grief, which is where wisdom is gained. This descent is not usually perceived as a step into power; it is usually felt as loss. But Persephone's descent (and it may be so for any being) is to sovereignty; it is said traditionally that She "becomes Queen of the Underworld," as She immerses in the knowledge of the dark realm, the "Underworld," beneath the visible. It is a stepping into the power of wisdom, gained through loss. Power (sovereignty) in this Cosmology is understood as coming from the Deep (within all), familiarity with the Deep, from vision, not as an ability to "lord" it over others: that is "control."[613] This a good season for reflecting on power as

---

611 See De Shong, *Inanna Lady of Largest Heart.*

612 See Jules Cashford, *The Moon: Myth and Image* (London: Octopus Publishing Group Limited, 2003).

613 Judy Grahn describes the difference between *power* and *control* in "From Sacred Blood to the Curse and Beyond," *The Politics of Women's Spirituality*, 265. She says that only the apple tree has the power to grow apples; everything else done to the apples is about control. Power needs

an authority that is earned by shamanic capacity, which is a capacity to travel the depths of transformation, and a capacity of vision to see the Thread of Life that continues beneath the visible, the connection of the joy and grief.

Autumn Equinox is a time for grieving our many losses, as individuals, as a culture, as Earth-Gaia: for contemplating the costs involved in the depth of the moment. And it may be felt as rage/anger, and space may be made for expressing this in the ceremony: the "Grief Process" in the offered script in this chapter may be also inclusive of a "Rage process." I have called such a process as we did one year a "Holy Tantrum." At this time we may join Demeter, and any other Mother Creator Goddess from around the globe, in Her weeping for all that has been lost. The Mother weeps and rages, the Daughter leaves courageously, the Old One beckons with Her wisdom and promise of transformation; yet all three know each other deeply, and share the unfathomable grief. But Persephone as Seed represents the thread of life that never fades away. The revelation of the Seed, central to this seasonal celebration, is that:

> "Everything lost is found again,
> In a new form, In a new way.
> She changes everything She touches, and
> Everything She touches, changes." [614]

And so it will be. In this way Persephone as Seed, tends the sorrows, "wholes" the heart.

The metaphor arose at a time in the human story when the power of the seed was coming to be understood – thus it is often said poetically that Demeter gifted humans with agriculture. Demeter, as a Grain Goddess is an Earth-Mother figure, in the lineage of the Old European Pregnant Goddess of all European folklore, and as such is also Mother of the Dead.[615] Demeter may be identified with the triangle motif which

---

to be re-understood as an organic capacity within life, always surging for unfolding.

614 Starhawk, *The Spiral Dance*, 115.

615 Gimbutas, *The Language of the Goddess*, 141.

is a common motif in ancient images, "schematized configurations of the vulva and the pubic triangle," representing the sacred source of life.[616] As such She is all three aspects of Creativity, may be storied as "Great Three-in-One,"[617] an eco-Trinity. She may be entitled "Mother of the Great Triangle of Life … complete in all Her parts: Creator, Preserver, Destroyer."[618]

[Figure 28] Demeter/Mother Hands Persephone/Daughter-Self the Wheat

This image above has been central to my understanding of Demeter and Persephone over the decades, and thus to PaGaian Autumn

616 Ibid., 145.
617 Edwards, *The Storyteller's Goddess*, 178.
618 Ibid., 178.

Equinox celebrations.[619] I suggest having this image of sacred wisdom present to you all year round. Demeter is passing on the knowledge of life – it's emergence and death – the mystery, which is represented by the wheat. For Persephone, who receives Demeter's gift, it is an acknowledgement and an initiation: acknowledgement of Her maturity and readiness for power, and the active initiation into that, the gifting and receiving. The icon is about a complex reality:

• It is Demeter's gift of agriculture to the world – humans are becoming aware of the planting of the seed, and the awesome power of the seed, moving out of simply hunting and gathering. This domestication of seed occurred about ten thousand years ago, wheat and barley being the earliest.

• It is at the same time, Demeter's gift of the Mysteries – moving into death with awareness, with eyes open, in that sense, coming into power; passed from the Mother to Daughter, all the way back in unbroken thread to the original Mother. These Mysteries identify each initiate with the seed (the "Persephone") that falls from the dying plant to lie underground, hidden from life, and yet sprouts again – in some form, shape or other. We will all be lost – we have been lost. Everything lost is found again, in a new form, in a new way. We are the Beloved Ones who are lost, along with everything else.

• The wheat represents a harvest of life that we are given: we are given all that went before us, it is handed to us. We may rejoice in it and give thanks. But like the seed that goes into the Earth, every moment of life is lost – it dissolves, and is never repeated. Also it has had a cost. This moment and all that we enjoy, has had a cost. Every moment in our lives and in the entire history of Gaia-Universe is never repeated – it is lost. That means it is also ever-new. That is the Mystery – it is lost, but it is ever-new. There is a place on that edge of the grief and the joy, where the Universe hums in balance – a creative tension (some may call it the "curvature of space-time"). It is what enables Creativity to go on … in our lives and in the Cosmos. This continuity, the unbroken thread, may be represented in the red thread that initiates sometimes wore, representing the sacred Thread of Life that has never faded away: if it

---

[619] Image [Figure 28] credit: Hallie Iglehart Austen, *The Heart of the Goddess* (Berkeley: Wingbow Press, 1990), 73.

had we would not be here.

Autumn Equinox/Mabon Breath Meditation

To begin this Seasonal Moment's contemplation:

Centering for a moment … feeling the full cycle of your breath. With the inbreath, the light phase of the cycle, we may celebrate the Virgin/Young One aspect, the Urge to Be. With the outbreath, the dark phase of the cycle, we may celebrate the Old One aspect, She who creates the Space to Be. At the Solstices – with the peaking and full emptiness at the bottom of the breath, we may celebrate the Mother aspect, She who is this Dynamic Place of Being, the Sacred Interchange. At the Equinoxes, we may celebrate the Sacred Balance of all three – in relationship, that keeps it all spinning. Feel the full cycle of your breath – the delicate balance of all Three.

Recognising the Triple Goddess Dynamic – the Creatrix – here in this primary Place … in your breath.[620]

Some Autumn Equinox motifs

[Figure 29]

620 See Appendix J for a "Triple Goddess Breath Meditation," which I developed as an embodied summary of this whole PaGaian cosmology.

Suggested decorations for this Seasonal Moment – on the front door, windowsills, corners of shelves, and/or altars – may include:

- **purple –** it is the colour traditionally used for sovereignty since the dye was expensive to make and thus reserved for royalty: even so decreed by law in the Elizabethan era. It is also frequently the colour for mourning, chosen for Catholic Lenten rituals perhaps for these reasons. Autumn Equinox is about descent to wisdom/sovereignty.
- **red ribbon/thread** – the Eleusinian initiates apparently wore them:[621] it is the thread of life that continues beneath the visible.
- **triangular fabrics** – representing Demeter, the Great Triangle of Life, the Great Three-in-One, and the sacred Vulva, source of life.
- **wheat stalks** – which is what Persephone takes with Her to the Underworld, and the Harvest of Life handed to Her. It could be any grain stalks, native to the region.
- **flower bulb** – the Seed… its power and magic, and "redemption."
- **pot of soil** – representing the descent into the dark earth.
- **mask** – representing the balance of grief and joy, inseparable: and may also be worn by the celebrant for the invocation of Demeter.

An Invocation[622]

One day Demeter and Persephone were sitting on the slope of a high hill looking out in many directions over Demeter's fields of grain. Persephone lay on her back while her Mother stroked her hair idly.

> *Mother, sometimes in my wanderings … I have met the spirits of the dead hovering and drifting aimlessly. … I spoke to them Mother. They seem confused and many do not seem to understand their state. Is there no-one to receive them and comfort them? …*
> *The dead need us Mother. I will go to them.*

Demeter abruptly sat upright as a chill passed through Her and rustled the grass around them. She was speechless for a moment, but then hurriedly began recounting the pleasures they enjoyed in their world

---

[621] I don't remember the reference for that.

[622] This invocational story is primarily the words of Spretnak, *Lost Goddesses of Early Greece*, 111-114, with my own summary.

of sunshine, warmth, and fragrant flowers. She told her daughter of the dark gloom of the underworld and begged Her to reconsider.

Persephone sat up and hugged her Mother and rocked Her with silent tears. …

They stood and walked in silence down the slope towards the fields. Finally they stopped, surrounded by Demeter's grain, and shared weary smiles.

> *Very well. You are loving and giving and we cannot give only to Ourselves. I understand why You must go. Still, You are my daughter and for everyday that You remain in the underworld, I will mourn Your absence.*

Persephone gathered three poppies and three sheaves of wheat. Then Demeter led her to a long, deep chasm and produced a torch for her to carry. She stood and watched her daughter go down further and further …

Persephone held Her Mother's wheat close to her breast, while her other arm held the torch aloft. She was startled by the chill as She descended.

The Universe Story and Autumn Equinox

At Autumn Equinox when dark reaches a new level of power, and it is storied as the departure of "Persephone," the Beloved One, there is opportunity to recall and express all the grief of the losses involved in Gaia's penchant for change and "dis-mantling." In the long evolutionary story, there have been many told and untold losses – species of flora and fauna that will never arise again, cultural losses, genocides, individual tragedies: Autumn Equinox is a time for remembering both the rich harvest gained and apparent for many, and also this deep loss and pain. Another layer to this recognition of loss at this Seasonal Moment, is the loss of every moment of existence – the fact that every moment dissolves and is never repeated: the story of Gaia-Universe is irreversible and nonrepeatable.[623] Every moment is thus totally new. We may grieve the loss and celebrate the Moment. Equinox is also then the

---

623 As Swimme articulates in video 10, *Canticle to the Cosmos.*

time to celebrate the delicate balance, the Creative Curvature of Space-Time, that Creative Edge upon which all life proceeds: and the red thread with which the wheat is tied for this ceremony may represent that.[624] And the seed, the "Persephone" that is planted ceremonially, may represent that very perdurable balance and fecundity that has enabled the entire evolutionary story. Six months later, at Spring Equinox/Eostar, the flowered seed is held up, as evidence of Her never-ending presence and generativity.

Brian Swimme says in his *Canticle to the Cosmos* series: "We know that we will die. We know that species die. Now we are aware that our Sun and planets will evaporate. In five billion years there will be no more Earth. The difficulty of bearing that loss is commensurate with the magnitude of the promised achievement;"[625] and so it is in our personal lives. The hope then is in the overall arc of Creativity.[626] Equinox is a celebration of the sacred balance – the creative curve between the compost and the lotus … they do exist together: it is the creative tension in which all apparent life is born, the delicate balance that scientists have named as "the curvature of space-time," in which Creativity of the Universe is possible.[627] That is one of the layers that I see in the image of Demeter handing Persephone the wheat. It is what enables Creativity to go on, in our lives and in the Cosmos.

We celebrate this creative tension at both Equinoxes. The balance of the three faces of Goddess, the three aspects of Cosmogenesis may be described in the words of Brian Swimme and Thomas Berry as "a fecund balance of tensions."[628]

From Swimme and Berry, an evolutionary story of the Creative process of the dark:

---

624 In our current times the thread may also represent the superstrings conjectured by Western science to form the basis of matter: see Brian Greene, *The Elegant Universe* (London: Vintage, 2000).

625 Video 5, "Destruction and Loss."

626 As Thomas Berry notes in his *Principles of a Functional Cosmology*. See Appendix A, principle 5.

627 Brian Swimme has said that his faith is in the curvature of space-time.

628 *The Universe Story*, 54.

> If the atoms in the prestellar cloud had been given language and the power to reflect upon inner experience, so that they could ponder the significance of the density waves sweeping through them and the rush of atoms ramming up against them, they would even then not have been able to speak in clear terms about the star they were destined to become. ... The beauty of the star gripped the atoms in some primordial manner; the beauty of the new flowering of Earth's realities likewise grips us and is in many ways the central significance of all our experiences of obstacle, disappointment, dismay, and despair. ... We cannot know with certainty ... what is required of us now. We will find our way only with a deep and prolonged process of groping – considering with care a great variety of interpretations, weighing evidence from a spectrum of perspectives, attending with great patience to the inchoate, barely discernable glimmers that visit us in our more contemplative moments.[629]

Earth Herself may be understood as Seed, and Brian Swimme does describe Earth that way in his *Canticle to the Cosmos* series[630]. He likens all the planets of our solar system to seeds in the soil of solar energy, and speaks particularly about the dimensions of Earth's sacred balance, in the context of enabling Her fluid nature, contributing to Her self-organizing and magical capacities.

Starhawk has a very germinal Seed Meditation in *The Earth Path,*[631] wherein the seed is held and asked to show you its history and wisdom, all who have guarded this "chain of life" over the eons; and gratitude may be expressed to the seed, the ancestors, the land, and the elements. Starhawk includes suggesting a question addressed to the seed, of what it may need and desire in order to thrive. I suggest that any felt response to these questions and contemplations may be understood as a "commissioning": that is, feeling for the sense of one's role as Seed, one who participates in the restoration of beauty, the vision of the Seed.

---

629 Ibid., 58.

630 Program 6, "A Magical Planet."

631 San Francisco: HarperCollins, (2004), 174.

Notes on imagery and metaphors in the offered Ceremonial Script

For the Call to Gather, drums with a strong beat is good: it is another seasonal moment of stepping into power, reflecting that of Spring Equinox/Eostar, but at this time heralding dramatic descent.

At the Equinoxes the focus for the Centering/Breath Meditation is on the experience of balance of light and dark, as Earth is so balanced in this Moment; imagining the breathing in as a swelling with light, and the letting go of the breath as a letting go into dark. It is spoken as "a fertile balance of tensions" that at the Spring Equinox is visualised as about to tip into increased light, and at the Autumn Equinox is visualised as about to tip into deeper dark.

In the process of Creating the Sacred Space – Calling the Directions/Elements, the directions are referred to as 'Places' that all gathered come from. This is a way of recalling the ancient ceremony of the Eleusinian Mysteries upon which this ceremony is based, wherein it is said that people came to this annual event from every corner of the Earth to be initiated, and these Mysteries were thus thought to hold the entire human race together.[632] Referring to and calling the directions in this manner also alludes to the deep truth that we (the atoms of our bodyminds) do indeed come from everywhere and everytime, in the evolutionary story. It may recall our ancient and multifold initiations in Her. Also, it may recall that participants and their lineages do come from many places of the Earth in more recent times.

Included in this process there is also an affirmation that "She is alive in me, and I in Her." This echoes part of the Seasonal rite of Spring Equinox,[633] and in particular, it recalls the Seasonal theme of the continuity of life, as the Mother Demeter passes all knowledge to the Daughter-Self Persephone, who thus becomes Mother – they are in each other; and it recalls how we are embedded in Earth. The elements of Water, Fire, Earth and Air are thus affirmed as associated with this continuity of life within us. The rhythm of the calling of the elements may be quicker than usual, and the responses chanted, to create a sense of this continuity of life beneath the visible.

Regarding the Invocation: this is a rite of initiation into the dark

---

[632] Durdin-Robertson, *The Year of the Goddess*, 158.

[633] See Chapter 7.

mysteries, the heraic descent of the Courageous One, the Beloved Daughter-Self, who must know the processes of the depths. Demeter is invoked with a chant, and She hands to each participant three stalks of wheat tied with red ribbon, signifying the passing on of the sacred knowledge of life.[634] The Mother is acknowledging each participant's readiness for the Journey, wherein they will receive and perceive the wisdom necessary for life to continue: in fact, with the gift of the wheat, She hands each one all that is needed. We are the bearers of life as much as we are borne by life.

Regarding the words with which Demeter addresses each participant/initiate: at Eleusis, all the participating celebrants identified themselves with Demeter. "Men who took part were given names with feminine endings. All the mystae wore the same clothes, simple robes …;"[635] men were included in acting out the myth of loss and recovery. "All people became the Mother in the Mysteries. They may also have identified powerfully with Persephone … ."[636] In the offered script, the Mother is invoked in all, identifying all as Gaian beings – women, men and all genders. She with whom all participants identify is the Mother of all life; and all identify also as Persephone/Daughter/Child,[637] sacred receiver and carrier of the Thread.

At Autumn Equinox, which is the main thanksgiving harvest festival in this Cosmology, the Communion rite is a sharing of food which each participant has brought and presented earlier in the ceremony as a representation of their abundance gained, their "harvest."

As noted above the suggested use of a triangular altar, and

---

634 This "knowledge" that the wheat signifies is as much genetic, body memory, as much as it is mind, trans-genetic cultural memory. This passing on of knowledge was originally understood as particularly expressed in the mother-daughter relationship: this turns out to have some basis genetically. See Coates, *The Seed-Bearers*. It may be understood to represent all that each being is given by the Universe in every moment – each one's harvest.

635 Pollack, *The Body of the Goddess*, 220.

636 Ibid., 221.

637 "Daughter" is understood as a title, not simply as sex of the child. Participants could choose "Child" or simply "Persephone."

triangular fabrics is part of the invocation of Demeter as the Great Triangle, Three-in-One.

The tying of red threads on each other is a confirmation of participation in the Vision of the Seed, the thread of life that continues beneath the visible; so that each may consider themselves commissioned by the Mother, and perhaps as Initiates into the Mysteries.

[Figure 30] An Autumn Equinox/Mabon Ceremonial Altar

[Figure 31] An Autumn Equinox/Mabon Ceremonial Underworld

AN AUTUMN EQUINOX/MABON CEREMONIAL SCRIPT

Note that the directions here are called in "counter-clockwise" direction which is sunwise in the Southern Hemisphere. The order may be changed to "clockwise"/sunwise for the Northern Hemisphere. Note that this particular offered ceremonial script is not considered prescriptive, and that there are many creative possible variations.

Some requirements and some suggestions: Wreath with wheat and Autumn leaves, around centre candle. Three wheat stalks tied with red ribbon for each participant – all wrapped in a special cloth and tied ceremoniously with long thick red ribbon. Purple triangle for altar cloth, overlaid on autumn coloured fabric perhaps. Basket of flower bulbs. Each bring food and/or drink to share, possibly harvest from their garden, placed next to themselves in circle. Glasses, plates, cutlery, bottle openers, napkins ready. Each bring a plant pot with soil – placed on a ceremoniously covered table in the "Underworld," which may be outside in a shed: this table also has a large pot full of soil, with small spades. Garden flame torches to light path to the Underworld. Each bring a shawl and stories of loss/grief. Candle or light near music switch. Tissues. Bell. Bowl of water and towel near "gate" of the "Underworld" for handwashing. Icon of Demeter and Persephone. Bowl of pomegranate seeds in Underworld. Red threads cut and ready for tying. Music ready for process of remembering losses.

Call to Gather … energetic drumming is good, until all are gathered, or a little longer.

Centering and Statement of Purpose

This is the Moment of the Autumnal Equinox, in the Southern Hemisphere – the moment of balance of light and dark in the dark part of the cycle. The light and dark parts of the day in the South and in the North of our planet, are of equal length at this time.

Feel the balance in this moment – Earth as She is poised in relationship with the Sun. See Her there in your mind's eye. Contemplate this balance. Feel for your own balance of light and dark within. Breathe into it. Breathe in the light, swell with it, let your breath go into the dark, rest with it. Feel for your centre, shift on your feet, from left to right, right to left ... breathe into it – this Sacred Balance, from which all comes forth.

In our part of Earth, the balance is tipping into the dark. Feel the shift within you, see in your mind's eye the descent ahead, the darkness growing, remember the coolness of it. This is the time when we give thanks for our harvests – the abundance we have reaped. And we remember too the losses involved. The story of Old tells us that Persephone, Beloved Daughter, receives from Her Mother, the wheat – the Mystery, knowledge of life and death – for this she gives thanks. She receives it graciously. But she sets forth into the darkness – both Mother and Daughter grieve that it is so.

Calling the Directions/Elements – Creating the Sacred Space

"Let us enter the sacred space, wherein we may speak the

Mystery. Let us begin by remembering from whence we come and who we are."

Celebrant: We come from the East to this Place, and we remember that we are Water. She is alive in us and we in Her.
(celebrant and all may dramatise the 'bringing in' with gesture of arms)
Co-celebrant carries water around and sprinkles it on participants.
Each responds as they feel the water, and begins to chant:
"I remember that I am Water. She is alive in me and I in Her." (it will produce an effect of a continuous thread of sound, as each starts the chant in turn).
Water carrier repeats as she puts the water down, and the chant stops: "I remember that I am Water. She is alive in me and I in Her."

Celebrant: We come from the North to this Place and we remember that we are Fire. She is alive in us and we in Her.
(celebrant and all may dramatise the 'bringing in' with gesture of arms)
Co-celebrant lights the fire, and takes it around the circle; each passes their hand over it.
Each responds as they feel the fire, and begins to chant:
"I remember that I am Fire. She is alive in me and I in Her." (it will produce an effect of a continuous thread of sound, as each starts the chant in turn).
Fire carrier repeats as she puts the firepot down, and the chant stops: "I remember that I am Fire. She is alive in me and I in Her."

Celebrant: We come from the West to this Place and we remember that we are Earth. She is alive in us and we in Her.
(celebrant and all may dramatise the 'bringing in' with gesture of arms)
Co-celebrant carries a rock or bowl of earth around the circle, for each to touch.
Each responds as they feel the earth, and begins to chant:
"I remember that I am Earth. She is alive in me, and I in Her." (it will produce an effect of a continuous thread of sound, as each starts the chant in turn).
Earth carrier repeats as she puts the earth down, and the chant stops: "I remember that I am Earth. She is alive in me and I in Her."

Celebrant: We come from the South to this Place and we remember that we are Air. She is alive in us and we in Her.
(celebrant and all may dramatise the 'bringing in' with gesture of arms)
Co-celebrant lights and carries the smudge around the circle, for each to smell.
Each responds as they smell the smudge, and begins to chant:
"I remember that I am Air. She is alive in me, and I in Her." (it will produce an effect of a continuous thread of sound, as each starts the chant in turn).
Smudge carrier repeats as she puts the smudge down, and the chant stops: "I remember that I am Air. She is alive in me and I in Her."

Celebrant: We have been present you and I (PACING THE CIRCLE) in the East, in the North, in the West, in the South, we have been present always in each stage of Gaia's story, as Water, Fire, Earth and Air – She is alive in us and we in Her. We are at Her Centre, which is everywhere. The circle is cast, we are between the worlds, beyond the bounds of space and time, where light and dark, birth and death, joy and sorrow meet as One.
(Celebrant light centre candle)

Invocation
Celebrant: Let us call Demeter, the Mother, who holds us, waits for us and receives us.
All chant – with drums: "Demeter, Demeter – Mother we call you, Mother we call you."
The celebrant proceeds to tie on the mask of Demeter, add to her costume and pick up the wrapped bundles of wheat, as the energy builds.

When "Demeter" is ready, She stands waiting: when the energy is built, the group finishes with proclamation of "Demeter!"

"Demeter" walks the circle ceremoniously (perhaps twice or more) untying the ribbon and unwrapping the cloth. She stands in place and holds up the unwrapped wheat bundles for all to see, stating:
"You are offered the wheat in every moment. It is given to you, if you will receive it."

"Demeter" goes to each one:
"(name), I give you the wheat – the Mystery – the knowledge of life and death. I let you go as Child, (Daughter/Mabon) most loved of mine: you descend to wisdom, to sovereignty. You will return as Mother, co-Creator with me. You are the Seed in the Fruit, becoming the Fruit in the Seed. Inner wisdom guides your path."

Each responds (with variation as desired for self expression): "It is so. I am Daughter (Child/Mabon/Young One), becoming Mother – Seed becoming Fruit. I am deepening into/descend to, wisdom, into sovereignty. The Mother knowledge grows within me."

When Demeter has given the wheat to all she (as celebrant) takes off her mask and extra "Demeter" garments, walks the circle counter-sunwise (to unwind), and retakes her own position in circle. A co-celebrant puts on the Demeter mask, and also other garments if desired, walks the circle sunwise as "Demeter" and gives the celebrant the wheat and blessing.

The co-celebrant then takes off the Demeter mask and other garments, walks the circle counter-sunwise back to her place in the circle.

Thanksgiving for the Harvest
Celebrant: Let us give thanks for our harvests – this harvest we are given in every moment, all that we have gained. Present your harvest, and speak if you wish.

Each picks up their "harvest" in turn (the food or drink they brought to circle), and puts it in the centre, speaking if they wish:
"I give thanks for ...."
Group response: "Your life, it is blessed." OR "Your harvest, it is wonderful. We hear your joy." (with variation as felt)

When all are done, celebrant concludes with: "We have harvested much, our lives are blessed. We are Daughters and Sons, Children of the Mother."

Remembering the Losses: Sorrow and Anger
Celebrant: It is time now to take up our wisdom (she picks up her wheat) and all that we have gained, and remember the sorrows, the losses involved. Let us put on our shawls and remember the grief and the rage of the Mother, of mothers and lovers everywhere, our grief and rage.

All put their shawls on, and may choose to put shawls over their head.

MUSIC ON[638]
Celebrant: Persephone descends. The Beloved One is lost.

The circle begins walking slowly, processing around, or each wandering within the ceremonial space.
Celebrant: Hold your wheat close to your breast, as we descend. Speak of that for which you grieve and rage.

Participants may speak names and events for whom and what they grieve, or rage: "I remember …" OR "I am angry about …"
Group response: "For this we grieve (or rage)." OR "We hear your anger (or grief)."

Repeat for as long as desired/felt.
Celebrant: Sit with a partner now and tell each other anything further in your heart, the losses for which you grieve and rage … or perhaps sit in silence and listen to the silence. Decide silently who will speak/express first.

Celebrant rings a bell after a few minutes as the cue to switch.

Celebrant rings the bell to bring all to finishing, and silence.

Song
Any co-celebrant in the circle may begin softly:
Ancient Mother we hear you calling, ancient Mother we hear your song.

---

638 I have used *Gentle Sorrow*, Dreams CD by Sky (1988). http://www.newagemusicbysky.com.

Ancient Mother our grief and pain is yours, ancient Mother we taste your tears.[639]

Silence

The Hope of Persephone
The basket of flower bulbs is passed around. Each takes one.

Celebrant: Persephone goes forth into the darkness to become Queen of that world. She tends the sorrows. These represent our Persephones, who tends the sorrows – we are the Persephones, who may tend the sorrows. Let us go out into the night with Her and plant our seeds.

Celebrant takes the lit lantern, and all follow her to the "Underworld," singing as they go:
"She changes everything She touches, and everything She touches, changes."[640]
Each plants their bulb in their pot, as the singing continues.

Story in the Underworld[641]
When all are done planting, a co-celebrant reads:
READER 1:

"Persephone had gathered three poppies and three sheaves of wheat. Then Demeter had led Her to a long, deep chasm and produced a torch for her to carry. She had stood and watched Her Daughter go down further and further into the cleft of the Earth. …

For months Persephone received and renewed the dead without ever resting or growing weary. All the while Her Mother remained disconsolate. … In Her sorrow She withdrew Her power from the crops, the trees, the plants. She forbade any new growth to blanket the Earth. The mortals planted their seed, but the fields remained barren. Demeter

---

639 This is a variation of a traditional Pagan/Goddess song. The author seems to be unknown.
640 This is part of the Kore Chant in Starhawk, *The Spiral Dance*, 114-115.
641 The story is an excerpt with slight alterations, from Spretnak, *Lost Goddesses of Early Greece*, 114-117.

was consumed with loneliness and finally settled on a bare hillside to gaze out at nothing from sunken eyes. For days and nights, weeks and months She sat waiting."

READER 2: (READER 1 move to back and put on mask)

"In the crook of Her arm Persephone held Her Mother's wheat close to Her breast, while Her other arm held the torch aloft. She was startled by the chill as She descended, but She was not afraid. Deeper and deeper into the darkness She continued, picking her way along the rocky path. For many hours She was surrounded only by silence. Gradually She became aware of a low moaning sound. It grew in intensity until She rounded a corner and entered an enormous cavern, where thousands of the dead milled about aimlessly, hugging themselves, shaking their heads, and moaning in despair.
(All may moan and dramatize this)

Persephone moved through the forms to a large, flat rock and ascended. She produced a stand for her torch, a vase for Her Mother's grain, and a large shallow bowl piled with pomegranate seeds, the food of the dead.
(READER 1 move to front of group to slightly elevated platform.)
As She stood before them, Her aura increased in brightness and in warmth.
READER 1: "I am Persephone and I have come to be your Queen. Each of you has left the body you knew and resides in the realm of the manifesting – the realm of the dead. If you come to Me, I will initiate you into your new world."
READER 2: She beckoned those nearest to step up onto the rock close to Her. As each one came before Her Persephone embraced them and then stepped back and gazed into the eyes. She reached for a few pomegranate seeds, squeezing them between Her fingers. She painted the forehead with a broad swatch of the red juice and slowly pronounced:
READER 1 (Persephone):
"You have waxed into the fullness of life,
And waned into darkness;
May you be renewed in tranquility and wisdom."

Each person comes forward to receive the red juice on their forehead and the blessing from Persephone.

Persephone (READER 1) takes off her mask, and another co-celebrant puts the mask on and blesses her as 'Persephone' has done for the others.

Moment of silence – then the group returns to singing:
"She changes everything She touches, Everything She touches, changes," as all move back to the circle and altar, carrying their pots with them.

One of the group may take the role of an attendant at the door/gateway to assist with holding pots while each rinses and dries their hands.

Communion

When all are back in circle, the celebrant holds up her pot, and pronounces:
"These represent our hope. The Seed of life never fades away. She is always present. Blessed be the Mother of all life. Blessed be the life that comes from Her and returns to Her."[642]

All may affirm this with response: "It is so."

Celebrant invites: "Let us eat, drink and enjoy Her gifts – the feast."

All join in serving and sharing the food and drink that has been presented by each in the circle.

Stories[643]

After all have been served and are settling into eating and drinking, the celebrant invites:
"Let us tell each other stories of the sacred balance – of grief and joy, the care that we feel and know; or anything else you would like to speak of."

---

642 The last two sentences of this pronouncement are from Starhawk, *The Spiral Dance*, 209.
643 This is a space for ceremonial storytelling: that is, with the practice of ceremonial manners.

Each speaker may be formally responded to by the group: that is, for example with "We hear you."

Red Threads

Celebrant invites: Let us tie red threads on each other. We participate in the *Vision of the Seed*, of the wheat, the *Thread of Life* that continues beneath the visible. We may consider ourselves initiates into the Mysteries.

The basket/bowl of red threads are passed around the circle, and participants tie them on each other's wrist/ankle.

Open the Circle

Celebrant invites all to open the circle. All may turn to the South:
We have remembered this evening that we are Air – present at each stage of Gaia's story, that She is alive in us and we in Her. May there be peace within us.
All: May there be peace within us. (each may gesture with folded arms over the heart or as they wish)

Celebrant (all turning to the West): We have remembered this evening that we are Earth – present at each stage of Gaia's story, that She is alive in us and we in Her. May there be peace within us.
All: May there be peace within us. (each may gesture with folded arms over the heart or as they wish)

Celebrant (all turning to the North): We have remembered this evening that we are Fire – present at each stage of Gaia's story, that She is alive in us and we in Her. May there be peace within us.
All: May there be peace within us. (each may gesture with folded arms over the heart or as they wish)

Celebrant (all turning to the East): "We have remembered this evening that we are Water – present at each stage of Gaia's story, that She is alive in us and we in Her. May there be peace within us."
All: May there be peace within us. (each may gesture with folded arms over the heart or as they wish)

Celebrant (all turning to the centre): We have remembered this evening that we are given the wheat, the mystery – the Mother knowledge grows within us. We are the seed in the fruit, becoming the fruit in the seed. We have remembered and given thanks for our abundant harvests, we have remembered the sorrows involved and we have remembered our hope, the sacred balance of the Cosmos – the thread of life, the seed that never fades away: it is the balance of grief and joy, the care that we may feel in our hearts. May there be peace within us and between us.
All: May there be peace within us and between us. (all join hands)

Celebrant: May the Peace and Care of Goddess go in our hearts and minds.

Song
The circle is open but unbroken. May the peace of the Goddess be ever in our hearts. Merry meet, and merry part, and merry meet again.
All: Blessed Be!

Points for individual contemplation prior to the ceremony:

- Your response to receiving the wheat from Demeter: what is the "wheat" … how you understand it at this time, what it is that you have been/are handed or offered by the Mother of All. Don't forget the basics of existence itself – these threads.
- Your "harvest" – all that you have gained: this is probably more personal, but still multi-dimensional. You may feel it as bigger – larger self, planetary.
- Losses for which you grieve and rage – personal, cultural, Gaian.

Autumn Equinox/Mabon Goddess Slideshow:
https://thegirlgod.com/pagaianresources.php

CHAPTER 12
CONCLUSION

This book is a documentation of a poetic process that has been practiced over some decades, and it remains in process, in myself and in other participants, and within any who pick it up in their own way. Her Creativity is never-ending, and we may consciously and creatively participate. The ancient ones who first built the stone circles may have done so for reasons intimately connected to sustenance/regeneration: that is, knowing what time it was in relation to gathering/growing of food, but that was not separate from the poetry, the cultural expression of Who and Where we are … and in these times such seamlessness may be renewed.

It is a whole yearlong process of practice, that will teach when done with consistency: that is, when not just choosing one or two Seasonal Moments of the whole cycle – though a nibble may give the taste of its power. The real learning will happen, She will teach, when the ceremonies are done devotedly: that is, religiously – as is common to say when something is done with intent and attention, understanding a thing's importance.

This particular documentation described in the previous chapters has been based on documentation made possible by my academic work, which included my ongoing and earlier experience. The academic doctoral work that I was so privileged to do, allowed organic and transpersonal research methods:[644] thus the time spent in ceremony, meditations, and intuitive cognitions was counted as part of the learning. This was deeply beneficial to my understandings of Her Poiesis.[645] I do feel/think that any documentation/reflection that any person does of a creative process, and certainly of this Creative process, will teach a new

---

[644] These methods are described to some extent in my book *PaGaian Cosmology,* 4-12, and in full in my doctoral thesis, *The Female Metaphor – Virgin, Mother, Crone – of the Dynamic Cosmological Unfolding: Her Embodiment in Seasonal Ritual as Catalyst for Personal and Cultural Change.*

[645] The evolution of some of the ceremonial scripts as I wrote them during the doctoral process is documented in the thesis noted above, Chapter 7.

depth to the experience.

The whole Wheel of the Year happens in every moment. The whole gestalt of the Wheel expresses Her extant Creativity. It is always Lammas, the sacred consuming; it is always Imbolc, the new emerging. Rebirth is always held in the compost: we live in a Universe that has revealed this. I learned the Poetry through practice of ceremony and with others, embedded it in my bodymind, made a start on the learning. "Poiesis" is the making of a world … and so, is the practice of heartfelt ceremony wherein Her eternal Creativity is acknowledged and expressed.

APPENDIX A

Thomas Berry's Twelve Principles of a Functional Cosmology

There is now a revised version of these, named as "Twelve Principles for Understanding the Universe" in *Evening Thoughts* by Thomas Berry (ed. Mary Evelyn Tucker), 145-147.

The version below, which is also Appendix A in the book *PaGaian Cosmology*, was taken from the Study Guide for the video series *Canticle to the Cosmos*. Thomas Berry's expression of these principles evolved over the years.

1. The universe, the solar system, and the planet Earth in themselves and in their evolutionary emergence constitute for the human community the primary revelation of that ultimate mystery whence all things emerge into being.

2. The universe is a unity, an interacting and genetically related community of beings bound together in an inseparable relationship in space and time. The unity of the planet Earth is especially clear; each being of the planet is profoundly implicated in the existence and functioning of every other being of the planet.

3. From its beginning the universe is a psychic as well as a physical reality.

4. The three basic laws of the universe at all levels of reality are differentiation, subjectivity, and communion. These laws identify the reality of the universe, the values of the universe, and the directions in which the universe is proceeding.

5. The universe has a violent as well as a harmonious aspect, but it is consistently creative in the larger arc of its development.

6. The Earth, within the solar system, is a self-emergent, self-propagating, self-nourishing, self-educating, self-governing, self-healing, self-fulfilling community. All particular life forms must integrate their functioning with this larger complex of mutually dependent Earth

systems.

7. Genetic coding process is the process through which the world of the living articulates itself and its being and its activities. The great wonder is the creative interaction of the multiple codings among themselves.

8. The human is that being in whom the universe activates, reflects upon, and celebrates itself in conscious self-awareness.

9. At the human level genetic coding mandates a further trans-genetic coding by which specific human qualities find expression. Cultural coding is carried on by educational processes.

10. The emergent process of the universe is irreversible and non-repeatable in the existing world order. The movement from non-life to life on the planet Earth is a one-time event. So, too, the movement from life to the human form of consciousness. So also the transition from the earlier to the later forms of human culture.

11. The historical sequence of cultural periods can be identified as the tribal-shamanic period, the neolithic-village period, the classical civilizational period, and the emerging Ecozoic era.

12. The main task of the immediate future is to assist in activating the inter-communion of all the living and non-living components of the Earth community in what can be considered the emerging Ecozoic era of Earth.

APPENDIX B

Gingerbread Snakes (for Samhain)

The recipe makes 28 snakes

Ingredients

5 cups wholemeal spelt flour (or mix of flours is fine)
4 tsp baking powder
1 tsp salt
4 tsp ginger powder and 1 tsp ground fresh ginger if you can
half cup butter
half cup oil
2/3 cup raw sugar (or brown)
1 cup molasses
2 eggs
half cup milk or soy milk or orange juice

Method

Cream butter and sugar
Add beaten eggs, oil, molasses, milk
Sift flours, salt, ginger, baking powder together
Add to wet ingredients
Chill overnight
Roll out to 1cm thick
Cut into snakes.
Put currants or little shiny balls, or sunflower seeds for eyes.
Bake at 180 C for 15 mins or until cooked looking.

APPENDIX C

Cosmic Walk Script (for Winter Solstice in particular)

I acknowledge Ruth Rosenhek, Miriam MacGillis, Lynn Margulis, Elisabet Sahtouris, Connie Barlow, Liz Connor, Brian Swimme and Thomas Berry in this composition. Note the times in more recent parts of the story will be different as years roll by.

From THE GREAT MYSTERY, which is wholly beyond all language and understanding – that Well of Creativity, Ultimate Reality which brings all things into existence, sustains all things, is revealed in all and to whom all returns …

*(1) from this great Mystery, some 13.7 billion years ago, is born the Universe … time, space and energy. THE GREAT RADIANCE. THE BIG O. This includes what we refer to as quantum fluctuation, inflation, expansion, gravity, electromagnetism, strong and weak nuclear forces, particle-antiparticle annihilation, creating a cosmos of matter and photons. Hydrogen and helium form.

*(2) A billion years later – 12.7 bya: PROTOGALACTIC CLOUDS of hydrogen form; the Universe differentiates into vast clumps of gaseous matter, forming a lacy pattern, like a never-ending web.

*(3) 11.7 bya: GALAXIES emerge. Gravity draws hydrogen into dense spheres of gas, sprinkled throughout each galaxy. At a threshold pressure, nuclear fusion begins – STARS are born, live, and die. Larger stars in their death throes explode and become supernovas blasting out into the cosmos. Supernovas are the mothers of the Universe, creating in their wombs the complex heavier elements that enrich the galaxies. Birth, death, and rebirth are ancient themes of the Universe.

*(4) 5 bya: birth of the SUN. Our Grandmother Star – Tiamat – huge in comparison to her child, our Mother Sun, becomes a supernova. In an explosion of possibilities She gives rise to our swirling solar system.

*(5) 4.6 bya – EARTH and the planets of our Solar system are born from aggregating debris in their orbital paths.

*(6) 4.5 bya: There is great bombardment – comets and meteorites pelt the Earth. The MOON is born when Earth is impacted by a Mars-sized body which also causes the Earth to tilt to the side giving rise to the seasons of the year.

*(7) 4.1 bya: the bombardment ends, Earth cools. Lightning storms rage. RAIN falls upon Earth for the first time – on and on. Great OCEANS form. Exuberant volcanoes expel hotly agitated deep earth to the surface. Over hundreds of millions of years, Earth has grown from dust particles to a large, hot, molten rock.

*(8) 3.8 bya: Bacteria emerge – possibly at great depth within Earth's crust or at hydrothermal fissures in the floor of the deep oceans. The FIRST LIVING CELLS ... the Mother cell – all our cells hold the memory of her. This Biosphere acts in association with geological and chemical activity to set up continuous cycles of Earth's finite reserves of elements vital for life: GAIA'S CLOSED CYCLE METABOLISMS are now in place.

*(9) 2.15 bya: PHOTOSYNTHESIS! Small creatures, who had run out of hydrogen from volcanic gases, learn to capture the Sun and use the energy, breaking apart H2O to feed on the hydrogen. Biomass increases as a result. A pollution crisis ensues as oxygen is left free floating in the environment, a waste product for these early life forms.

*(10) 2.1 bya: OXYGEN CRISIS threatens all life, when geology can no longer soak it up. Primordial cells unable to adapt to the new conditions disintegrate from the excess heat. But is also an opportunity because (i) an ultraviolet-absorbing OZONE SHIELD is formed in the upper atmosphere and (ii) RESPIRATION evolves, as a way to use the oxygen for high energy.

*(11) 2 bya: the EUKARYOTIC cell develops – individual organisms learn CO-OPERATION and specialization, evolving

symbiotic relationships. Life becomes NESTED, one centre of creativity within another, as oxygen using mitachondria and photosynthesizing plastids cooperate with larger host cells.

*(12) 1.5 bya: crisis conditions such as food shortages, lack of moisture or extreme temperatures drive hungry ancestral organisms to resort to eating each other – HETEROTROPHY. Sometimes, these tiny cellular beings cannot digest what they have devoured and a type of sexual union arises – MEIOTIC SEX. The genetic possibilities for life increase enormously.

*(13) 1.2 bya: the geological activity of MARS freezes up, foreclosing full geophysical cycling of chemical elements vital to life.

*(14) 560 million years ago: the first MULTICELLULAR organisms emerge – an innovation in which the offspring of dividing cells stay in bonded association with each other, resulting in the first communities and the birth of SYNERGY. It is a kind of self-organisation that allows many new kinds of cooperation and specialization for the good of the whole. There is more biodiversity. Organisms grow larger and more complex, and with less capacity for self-repair, resulting in DEATH BY AGING. Soon after there is also the FIRST MAJOR EXTINCTION (Cambrian) – was its cause glaciation, predation or combination?

*(15) 520 mya: SIGHT is invented, in various forms including sophisticated crystal lenses of calcite in the compound eyes of trilobite arthropods. Earth sees Herself for the first time.

The first soft-bodied animals evolve in the oceans. Over the next millions of years, animals invent hard parts for teeth, beaks and claws, and shells for protection.

*(16) 500 mya: the ANCESTRAL FLINDERS RANGES in Australia, and many mountains elsewhere are squeezed and folded into being. Supercontinents continue to fragment and reform, driven by radioactivity deep within Earth. Associated earthquakes and volcanic activity cause major cyclical climate changes.

*(17) 440 mya: Life ventures onto LAND. Algae and fungi symbiotically pool their talents, merging into the first LAND PLANTS. The continents grow green with low-lying ancestors of mosses. Leaving the water, animals such as worms and mollusks and crustaceans seek the adventure of weather and gravity. INSECTS evolve.

*(18) 400 mya: ANCESTRAL SPIDERS and the first AMPHIBIANS. Animals hop and lumber onto land, trading in their gill slits for air-breathing lungs, transforming fins into stubby legs and continuing to return to the water to lay their eggs.

The formation of the limestone beds that later become the Jenolan Caves after the rocks have been uplifted. Small scale mountain building is going on with lava flows, heavy erosion and upheavals.

And soon the SECOND MAJOR EXTINCTION (Devonian) – affecting marine creatures with carbonate shells.

*(19) 360 mya: the invention of the wood cell – the first TREES, as plants rise to new heights. The first subtropical FORESTS evolve. Over generations, these forests load themselves with carbon extracted from the atmosphere which later becomes fossilized as coal and oil.

*(20) by 320 mya: Earth learns to fly, as insects evolve FLIGHT. Earth learns to HEAR, as ancestors of the modern frog evolve the first vertebrate ears – hearing the first sound waves transmitted through air. The first REPTILES appear, with the first land worthy eggs that can survive out of water – the AMNIOTIC EGG with a shell and membrane. Reptiles also invent the penis and early version of vagina – copulation outside of water.

*(21) 250 mya: PANGAEA forms – a single supercontinent – in a series of steps over millions of years. The THIRD MASS EXTINCTION (Permian) and perhaps the most devastating (75 – 95 % of all species eliminated) – marking the end of the Paleozoic Era … possibly due to land and ocean re-arrangements, and resultant climate changes, and/or possible asteroid impact.

*(22) 230 mya: emergence of DINOSAURS. For millions of years these creatures flourish. Dinosaurs, sometimes as large as 40 meters, are social animals that often travel and hunt in groups. Dinosaurs develop a behavioral novelty unknown previously in the reptilian world – PARENTAL CARE. Dinosaurs carefully bury their eggs and stay with the young after they hatch, nurturing them toward independence.

PANGAEA BREAKS UP into Laurasia and Gondwana. Earth will bring forth six times more biodiversity as the continents draw apart. The PACIFIC and then the ATLANTIC OCEANS form. The FOURTH MAJOR MASS EXTINCTION (Triassic).

*(23) by 200mya: the FIRST MAMMALS, small and nocturnal, jump, climb, swing, and swim through a world of giants. Some rodent-sized insect-eaters evolve lactation, enabling mothers to spend more time in the nest keeping their young both fed and warm. Wind-dispersed POLLENS.

*(24) 150 mya: BIRDS emerge, a direct descendant of the dinosaur as leg bones evolve into wing bones, jawbones into beaks and scales into feathers. Far larger than today's birds, wing spans are as large as 12 metres.

*(25) by 114 mya: FLOWERS become an important part of the flora, evolving gorgeous and overt sexual organs, colors, perfumes, and delightful nectars. Insects are attracted and transport pollen from one flower to the next, fertilizing the plants on which they feed. MARSUPIAL MAMMALS – kangaroo and platypus have evolved. Now PLACENTAL MAMMALS too.

*(26) 65 mya : the Cenozoic Era begins, after the FIFTH MAJOR MASS EXTINCTION (Crustaceous), caused when a mountain size asteroid hits the Yucatan peninsula. This results in huge tidal waves, a magnitude 12 earthquake, acid rain, dust blocking light from the Sun, a firestorm that incinerates a quarter of the biomass releasing huge amounts of carbon dioxide that increases global temperature for a million years. Dinosaurs disappear, after 160 million years.

At the beginning of the Cenozoic Era – 65 mya – mammals embark on a fast evolution in the space created.

ULURU in Central Australia emerges.

*(27) by 30 mya: Antarctica has separated from Australia and South America giving birth to the Antarctic Circumpolar Current, changing the circulation of water around all the continents. The combination of this and Earth's now relatively stable tilt, leads to the formation of distinct CLIMATIC ZONES, with intensely different SEASONS.

FRUIT TREE FAMILIES originate. NUT TREES co-evolve with squirrels. Whale ancestors return to the sea. Bear families send ancestors of seals into the sea. There are monkeys, then apes.

*(28) 23 mya: GRASSES flourish – superbly adapted to survive mammal grazing and drought. Monkeys and apes split. Soon deer, later gibbons and orangutans, and still later gorillas, cats and dogs.

Australia is developing a distinct flora – with many rainforests. Conifers, including the Wollemi and Huon pines are abundant in places.

*(29) 5 mya: Chimpanzees and hominids. Within a million years the hominids leave the forest, stand up, and walk on two legs. The savannah offers the challenges and opportunities for these early ancestors.

Australian SOILS are laid down – some of the oldest on Earth, yet also short on some nutrients resulting in wide diversity as plants adapt.

*(30) 2.5 mya: the FIRST HUMANS (homo habilis) are making tools and possibly what we may call "art" – early symbolic representation. The light photons from Andromeda galaxy reaching Earth now in this moment began their journey through the Womb of Space.

*(31) 1.4 mya: Humans (homo erectus) domesticate FIRE – enabling migration to colder places, cooking, storytelling … a new multivalent power.

*(32) up to 500,000 ago: in this period humans have begun carving female figurines – ANCIENT MOTHER FIGURES.

SYMBOLIC LANGUAGE emerges. Clothing and shelter. Brown bears, wolves, llamas, archaic homo sapiens, soon cave bears, goats and modern cattle.

*(33) 100,000 ago: RITUAL burials – SHAMANIC and GODDESS RELIGIONS and ART are emerging. Wooly mammoths and wildcats have emerged and soon polar bears.

*(34) by 50,000 years ago: humans have entered AUSTRALIA. They and the dingo they bring with them, cause an extinction of the largest marsupials, reptiles and flightless birds – as was characteristic of almost all human migration. Human migration around the planet marks the beginning of the current SIXTH MAJOR MASS EXTINCTION.

Soon the Darug people to the east of the Blue Mountains and the Gundungurra people of the Western ridges hold intertribal ceremonies on the narrow main ridge of the upper and the lower mountains.

*(35) by 11,000 years ago: there are musical instruments, cave paintings. TAMING of dogs and soon sheep and goats, then cattle. AGRICULTURE and the DOMESTICATION OF PLANTS all over the globe emerges.

*(36) by 3,000 years ago: Cities have grown up. NEWGRANGE in Ireland has been built, the bluestones at STONEHENGE are in place. The Sumerian civilization has invented the WHEEL and CUNEIFORM WRITING. Soon the building of the Egyptian PYRAMIDS begins, and CLASSICAL RELIGIONS emerge – Judaism, Buddhism, Hinduism, Christianity, Islam. Rule of the Father emerges. There is CHRONIC WARFARE. Soon the earliest origins of the ALPHABET, and the Sanskrit language. Soon the time of Confucius, early Greek philosophers. Soon coinage as MONEY is invented.

*(37) by 450 years ago: the COPERNICAN revolution in Western consciousness, and the beginning of the mechanistic paradigm as primary metaphor for understanding the nature of Earth and the

Cosmos.

Large scale Western COLONIZATION of other lands has begun. The ROMAN INQUSITION has been put in place – persecuting and murdering humans of Earth-based religious practice.

*(38) by 100 years ago: the THEORY OF RELATIVITY is introduced – quantum physics, nuclear weapons and nuclear medicine. Western science gathers evidence of a DEVELOPING and EXPANDING UNIVERSE – distant galaxies and deep space are coming into human awareness.

*(39) by 60 years ago (at the time of writing this): Rachel Carson's Silent Spring begins new ecological awareness, DNA has been identified, the first MICROCHIP developed, a CHARTER OF HUMAN RIGHTS is adopted, subjugated races and peoples voice dissent and achieve change. BACKGROUND RADIATION – echoes of the Origins/Birth of the Universe – are detected.

*(40) Today (last few decades of 20th – early 21st centuries): humans have seen EARTH AS A WHOLE (from space), within the Womb of Space. Earth is again recognized as a living organism (not a machine or inert) – "GAIA." There is growing recognition that humans – with a tremendously expanded population and huge demands on the body of Earth – are causing the SIXTH MAJOR MASS EXTINCTION. PERMACULTURE is developed. The World Wide Web (INTERNET) is created. Humans are understanding the deep communion, relationship and subjective nature of all: realizing (perhaps again or anew) that we are IN the unfolding Story of the Cosmos, participating with our consciousness, exploring our role in this sacred awesome Event.

We may celebrate this sacredness, and the role human consciousness may play, we may enjoy the beauty of Earth and the Gift of it all.

Offered Conclusions for Cosmic Walk Ceremony

When the "Today" candle has been lit, and the narration concluded, the walker who has been lighting a candle at each point, lights an extra votive candle that has been left at the end, puts the carrying candle down, and makes a personal statement of presence such as "I am .... and I am here" or "I am .... and this story is my story."

The other participants are then each given a votive candle and invited to spiral their way to the centre. Each lights their candle from the Centre (the first person should take the taper lighting candle with them), and walks meditatively back out, placing their candle near the walker's candle at the end, and making some statement of presence if they wish.

AND/OR

EACH TURN TO A PARTNER. Narrator speaks: *Look at each other and see that the person you have before you recapitulates the entire story of the Universe – if it wasn't so they would not be here. Every cell remembers the first cell, the transformations all the way along, the body form is inherited from early creatures, the eyes carry the memory of photosynthesis .... and so on. Take turns at giving this vision to each other – and receiving it. Perhaps use gesture, perhaps speaking: "Here is the Source of everything, here is the Mystery of the Universe" – however you like. Thank each other.*

OPEN CIRCLE with song and percussion.

For more on the Cosmic Walk: https://pagaian.org/articles/cosmic-walk-script/.

APPENDIX D

Instructions for Cosmogenesis Dance
traditionally done at Winter Solstice but good anytime.

Done to the music "Adoramus Te Domine" or more recently to the music of "Devi Prayer: Hymn to the Divine Mother": see https://pagaian.org/book/appendix-i/ for more.

Start with a partner, decide who are #1's and #2's.

#1's form the inner circle with their arms raised, holding hands.
This has been known as the Stillpoint of the dance, and as such it has been regarded as more significant than the other two layers. I like to weight its significance as the same as the other layers, as is the case for the three aspects of Cosmogenesis. This layer then may be identified with autopoeisis, the centre of creativity that each being is.

#2's are eight steps behind their partner in an outer circle, and take one step to right before starting the dance, not holding hands.

On the second measure, the inner circle takes four steps backward, slowly lowering arms, and at the SAME time, the outer circle takes four steps forward and reaches in front of the inner circle people to grasp the hands of others in the outer circle: this forms an interwoven basket weave. This layer may be identified with the communion of being.

On the Adoramus Te of the second measure, when the circles are interwoven, everyone sways to the left first, the to the right. Then #1's drop hands – #2's continue to hold hands. #1's take four steps backwards, #2's take four steps forward holding hands and raising their arms slowly. On the next Adoramus Te, the 2's (inner circle) are standing still holding the Stillpoint. The outer circle people are swaying (to the left first), not holding hands. This outer layer is the form of unique individuals. It may be identified with differentiation.

NOTE: It is important to start the dance (walking the initial 4 steps) on the second measure: that is, start walk after 1st "Adoramus Te Domine" (if using that music) ... so that the dance finishes with all interwoven in the communion layer – which our Place of Being is essentially.

APPENDIX E

Silent Night (Cosmic version)
by Connie Barlow

NOTE: This "stardust" version of Silent Night was first used by Connie Barlow on December 21, 2003 at the intergenerational Sunday service of the Unitarian Universalist Fellowship of Clemson, South Carolina, USA.

For each verse, the first and last lines are always the same:

FIRST LINE:
Silent night. Holy night. All is calm. All is bright.
LAST LINE:
Life abounds upon Earth. Life abounds upon Earth.

1. PLAN-ets GRACE-fully CIR-cle the sun;
STAR-dust CY-cles through EV-er-y one.

2. RA-diant BEAMS from PRI-mor-dial stars,
CLUMPED into PLAN-ets like VE-nus and Mars.

3. CAR-bon, NI-tro-gen, and CAL-ci-um
ALL were BORN inside AN-ces-tral suns.

4. DEATH and re-CY-cling of MILL-ions of stars
BROUGHT forth PLAN-ets and ALL that we are.

5. SIL-ver, GOLD, and TI-ta-ni-um
FORGED in STARS before EARTH had begun.

6. FLAR-ing FORTH across HEAV-en above,
SU-per-NO-vas made ALL that we love.

For more detail and story: https://pagaian.org/2019/06/02/cosmic-silent-night-for-winter-solstice/.

APPENDIX F

PaGaian Joy to the World
Sung to the tune of "Joy to the World." This is an adaptation by Glenys Livingstone, 1998.

Joy to the World
The light returns
Let all receive Her Love

CHORUS:
Let every heart
Let every tongue
Repeat the sounding Joy
Repeat the sounding Joy
Let creatures, and all of nature sing

She moves the stars
With Her Desire
Let all receive Her Power

She grows the seed
With all Her Love
Let all receive Her Wisdom

She lights our hearts
She grows our food
Let all receive Her Joy

Joy to the World
The light returns
Let all receive Her Love.

Ref: Livingstone, *PaGaian Cosmology*, 307. In the original version the word "might" was used in the second verse: but upon further thought, I preferred the metaphor of Desire and it is more accurate.

APPENDIX G

Beltaine Yoni-Cake Recipe

Ingredients
2 and 1/4 cups wholemeal flour (spelt or gluten-free option)
1 and 1/2 tsp baking powder
1/2 tsp baking soda
1/2 tsp salt
1tsp vanilla
1/2 cup brown sugar
3/4 cup butter (or oil/butter combo)
2 eggs
1 cup buttermilk
cochineal for pink colouring

Method
Preheat oven to 350 F (180 C)
Sift flour with baking powder and baking soda and salt.
Cream sugar and butter. Add vanilla.
Separate eggs. Whip egg whites until stiff. Beat egg yolks and add to sugar-butter mix. Add sifted ingredients to this mixture, a third at a time, alternating with buttermilk. Add cochineal – sufficient to make good pink colour.
Fold in egg whites and turn into a well-greased and floured ring pan.
Bake 30 – 40 mins.

Cool slightly before turning out of pan.

At this stage you might want to freeze it until the day before you need it.

Then on the room-temperature cake, pour over a warmed-up mixture of honey and rose water, about half and half. I usually pour over quite a lot and let the cake soak in it a bit, before removing it to a fresh plate. Pour over a little more honey if you like. Cut it into slices but leave it whole on the plate. Stick pink/red/colour mixture rose petals on it and in the centre of the ring. To protect any cover you put on top of the cake at this stage, and to preserve the cake's appearance, put toothpicks into the top of it before covering.

[Figure 32] Beltaine Yoni-Cake

APPENDIX H

EARTH'S DESIRE
by Thomas Berry

To be seen
in her loveliness

To be tasted
in her delicious fruits

To be listened to
in her teaching

To be endured
in the severity
of her discipline

To be experienced
as the maternal source
whence we come
the destiny
to whom we return.

APPENDIX I

Salt Dough Recipe (for Lammas bread figures)
This is a summary version of a recipe by Janet Wood (used with her permission), which has more information about salt dough and creative possibilities. See http://www.ancientnile.co.uk/saltdough.php.

Ingredients:
2 cups plain flour
1 cup table salt
1 cup water

Optional:
1 tbsp vegetable oil (makes mixture easier to knead)
1 tbsp lemon juice (makes finished figure harder)

Method:
Mix flour salt and any of the optional ingredients in a large bowl, and gradually add water to soft dough consistency. Add more flour if too sticky, or more water if too dry.
When mixed place the dough on a flat surface that is coated lightly with flour and knead for ten minutes. Let the dough stand for twenty minutes before beginning to shape, or you can store it in the fridge in an airtight container for up to a week.

Drying:
This can be done naturally in open air, or with baking in an oven. The oven should be no hotter than 100C, and it is best to start at 50C and increase the heat after thirty minutes. Drying time will vary according to size and thickness, but on average natural drying takes thirty to forty-eight hours, and oven drying takes three to four hours.

I usually make these figures days ahead of Lammas ceremony and freeze them until the day before.

APPENDIX J

TRIPLE GODDESS BREATH MEDITATION
each Being as a Creative Place of Cosmological Unfolding

(done with movements of the Yoga Mudra)

With every inbreath I receive the Gift – I am Virgin ever-new.

With the peaking of every breath I am this dynamic Place of Being. I am Mother.

With every outbreath I become the Gift, the Old One – ever transformed.

(And because Virgin and Old One are often hard to distinguish … the clarity of the beginning and the end is not always apparent: the snake bites Her tail.)

With every breath in, I receive the Gift of All – the Transformations of the Ages, Old One.

With the peaking of every breath, I am the Sentient Place of Now, the Sacred Interchange, Mother.

With every breath out, I am the Gift to All – again renewed, Virgin.

(And very simply – and best done with expressive arm movements as suggested in the Turas Experience in Chapter 2)

With every breath I celebrate

- this particular beautiful Self, new in every moment
- in deep relationship and communion with Other, the web of Life
- directly participating in the Sentience of the Creative Cosmos, the Well of Creativity.

With every breath (with same arm movements as above):

I am the Ancient One,
ever-New,
in this Place.

For a YouTube version:
https://thegirlgod.com/pagaianresources.php

APPENDIX K

MoonCourt, PaGaian Ceremonial Space
Blue Mountains, Australia
Darug and Gundungurra Country

[Figure 33] MoonCourt

For some story about MoonCourt see:
https://www.magoism.net/2018/01/essay-mooncourt-goddess-ceremonial-space-by-glenys-livingstone-ph-d/.

For some story of its building see:
https://pagaian.org/articles/mooncourt/.

ACKNOWLEDGEMENTS

There have been so many essential contributions to this work, and compost aplenty amongst it. It does take the whole universe story to get to any moment, and so it is with any life, and creative work.

This work leans heavily in its origins on Starhawk's work, particularly her book *The Spiral Dance*. I was privileged to have attended Starhawk's first Reclaiming class series in San Francisco in 1980, and I learned so much from her over the years. The teaching of cosmologist/physicist Brian Thomas Swimme has been a major influence; all his videos and books enlarging my perceptions of *Who we are* and *Where we are*. The rich teachings of geologian and cultural historian Thomas Berry were very much part of that. Charlene Spretnak's restorying of Goddess in her book *Lost Goddesses of Early Greece* in particular, and in her articulation of Gaian spirituality in various articles, books and teachings was essential; and in her reading and editing of my doctoral thesis. There was Adrienne Rich's book *Of Woman Born* that taught me so much as I set out from the confines of patriarchal blinkers, the radical writings of philosopher Mary Daly, the sharp and poignant poetry of Robin Morgan, the medicinal mythologizing of Jean Houston, the solid foundation and inspiration of the work of archaeologist Marija Gimbutas. And I deeply thank the many, many other women whose books, teachings, research, art, poetry and song have nourished and informed me, helped bring me home; many are referred to in this book, and many more are simply woven into the fabric of my being.

I learned so much from hours of listening to tapes in the library of the Australian Transpersonal Association – Joan Halifax, David Bohm, Christina and Stan Grof, Jürgen Kremer, Rupert Sheldrake to name a few. My grounding in liturgical practice primarily with the Jesuit School of Theology in Berkeley as part of my Master's degree in theology and philosophy, developed my poetic sense and expression for the sacred. I thank my late friend Ian Brown who spoke by heart the poetry of Rilke, and much of Rumi and Iraqi, with female metaphor, whilst we walked and for gatherings; this magic lured me into the spelling effects of poetry.

For over two decades at the time of this writing, this work has clearly been supported and encouraged by my beloved partner Robert (Taffy) Seaborne. He in fact collaborated, especially with the building and maintaining of the ceremonial space MoonCourt, wherein many gathered and the poiesis could grow (see Appendix K). MoonCourt was his "Taf Mahal" he joked: that is, he built it for love. As told in the Preface, Taffy coined the term "PaGaian" itself, in the midst of a discussion wherein I was searching for placing my sacred practice as either Pagan or Gaian. His support is ongoing in this new phase of our lives.

I feel deeply thankful to Rob Blake who also coined the term "pagaian" from afar at a similar time, registered the domain name pagaian.org, then found me and my book *PaGaian Cosmology*, and gave me the website.

I thank all the people who came to ceremonies, events and classes, some of whom formed a consistent core over many years; it could not have happened without them, their desire for the story, their participation as co-celebrants, and quite often helping in the domestic preparation of the ceremonial space.

I thank Dr. Helen Hye-Sook Hwang deeply for her encouragement over years, with her invitation to me to write regularly for Return to Mago e-magazine, and for the honour of travelling with her to Korea as co-facilitator of the 2014 Mago pilgrimage. I thank Helen for calling this book forth and for much generous editing work and advice.

I thank Trista Hendren of Girl God Books for her enthusiastic response to my stories, and for the publications I have been part of in particular: and for her embrace of women and our passionate work for the world. I thank Trista for her kindness always, and for midwifing the final stages of the birth of this book.

I thank Kaalii Cargill for her kind editing help.

Glenys Livingstone Ph.D. (Social Ecology) has been on a Goddess path since 1979. She is the author of *PaGaian Cosmology: Re-inventing Earth-based Goddess Religion*, which fuses the indigenous traditions of Old Europe with scientific theory, feminism, and a poetic relationship with place. She was born and lives in country Australia, where she has facilitated Seasonal ceremony for decades, taught classes, and mentored apprentices. Her new book *A Poiesis of the Creative Cosmos: Celebrating Her within PaGaian Sacred Ceremony* documents the synthesis of her work over the past decades. She is the author of the children's book *My Name is Medusa*, and co-editor of the anthology *Re-visioning Medusa: from Monster to Divine Wisdom.* Glenys teaches a year-long on-line course "Celebrating Cosmogenesis in the Wheel of the Year" for both hemispheres. Her website is http://pagaian.org.

ADDITIONAL TITLES BY GLENYS LIVINGSTONE PH.D.

PaGaian Cosmology: Re-inventing Earth-based Goddess Religion

Re-visioning Medusa: from Monster to Divine Wisdom – co-edited with Trista Hendren and Pat Daly

My Name is Medusa – illustrated by Arna Baartz

Glenys has also contributed to the following anthologies:

Foremothers of Women's Spirituality: Elders and Visionaries, Miriam Robbins Dexter and Vicki Noble (editors), Teneo Press Inc., 2015. Essay title: "Conceiving and Nurturing a Poiesis of Her: a PaGaian Cosmology".

Goddesses in Myth, History and Culture, Mary Ann Beavis and Helen Hye-Sook Hwang (editors), Mago Books 2018. Essay title: "Goddess, Science and Paganism: a PaGaian Cosmology".

Inanna's Ascent: Reclaiming Female Power, Trista Hendren, Tamara Albanna and Pat Daly (editors), Girl God Books, 2018: with two essays.

Celebrating Seasons of the Goddess, Dr. Helen Hye-Sook Hwang & Dr. Mary Ann Beavis (editors), Mago Books, 2017: with seven essays/titles.

She Rises: How Goddess Feminism, Activism and Spirituality?, Helen Hye-Sook Hwang, Mary Ann Beavis and Nicole Shaw (editors), Mago Books, 2016. Essay title: "Celebrating Her/My/Our Everyday Sacred Journey Around Sun".

Jesus, Muhammed and the Goddess, Trista Hendren, Pat Daly Noor-un-nisa Gretasdottir (editors), Girl God Books, 2016. Essay title: "Exodus 1980 Revisited".

Godless Paganism: voices of Non-Theistic Pagans. John Halstead (editor), Lulu.com, 2016. Essay titles: "A PaGaian Perspective" and "A Poetry of Place".

She Rises: Why Goddess Feminism, Activism, and Spirituality, Helen Hye-Sook Hwang and Kaalii Cargill (editors), Mago Books, 2015. Essay title: "Now recognising Her in Me".

She is Everywhere Vol. 3, Mary Saracino and Mary Beth Moser (editors), Bloomington: iUniverse Inc., 2012. Essay title: "Spelling and Re-Creating Her".

Goddesses in World Cultures, Patricia Monaghan (editor), Praeger Publishers 2010. Essay title: "GAIA: Dynamic, Diverse, Source and Place of Being".

Indian Journal of Ecocriticism, Volume 3 August 2010. Essay title: "Female Metaphor, Science and Paganism: A Cosmic Eco-Trinity".

Songs of Solstice: Goddess Carols, Trista Hendren, Sharon Smith and Pat Daly (editors), Girl God Books, 2022. Essay title: "Winter Solstice as it is Told for PaGaian Ceremony" and carol: "PaGaian Joy to the World".

BIBLIOGRAPHY

Abram, David. *The Spell of the Sensuous.* New York: Vintage Books, 1997.

Adler, Rachel. "A Mother in Israel." In *Beyond Androcentrism: Essays on Women and Religion,* edited by Rita Gross, 237-255. Montana: Scholar's Press, 1977.

Anderson, William. *Green Man: The Archetype of our Oneness with the Earth.* Helhoughton FAKENHAM:—— COMPASSbooks, 1998.

Ardinger, Barbara. *A Woman's Book of Rituals and Celebrations.* New World Library, 1995.

Ashe, Geoffrey. *The Virgin.* New York: Arkana, 1988.

Austen, Hallie Iglehart. *The Heart of the Goddess.* Berkeley: Wingbow Press, 1990.

Barlow, Connie, ed. *From Gaia to Selfish Genes: selected writings in the Life Sciences.* Massachusetts: MIT Press, 1994.

__________. *Green Space, Green Time: the way of science.* New York: Springer-Verlag, 1997.

Baring, Anne, and Cashford, Jules. *The Myth of the Goddess: Evolution of an Image.* Penguin Group, 1993.

Berger, Pamela. *The Goddess Obscured.* Boston: Beacon Press, 1985.

Berry, Thomas. *The Dream of the Earth.* San Francisco: Sierra Club Books, 1990.

__________. *Evening Thoughts: Reflecting on Earth as Sacred Community,* edited by Mary Evelyn Tucker. San Francisco: Sierra Club Books, 2006.

Brandon-Evans, Tira. "Tailtiu: Harvest Goddess. In *Goddess Alive!,* issue 18, Autumn/Winter 2010: 10-12.

Brennan, Martin. *The Stones of Time: Calendars, Sundials, and Stone Chambers of Ancient Ireland.* Rochester Vermont: Inner Traditions, 1994.

Campbell, Joseph. *The Power of Myth with Bill Moyers.* New York: Doubleday, 1988.

Cashford, Jules. *The Moon: Myth and Image.* London: Octopus Publishing Group Limited, 2003.

Chicago, Judy. *The Dinner Party.* Hammondsworth: Penguin, 1996.

Chittick, William C. and Wilson, Peter Lamborn, trans. *Fakhruddin 'Iraqi: Divine Flashes.* London: SPCK, 1982.

Christ, Carol P., and Plaskow, Judith, eds. *Weaving the Visons: New Patterns in Feminist Spirituality.* New York: HarperCollins, 1989.

Cixous, Hélène. "The Laugh of the Medusa," translated by Keith Cohen and Paula Cohen. *Signs,* Vol 1 no. 4, Summer 1976: 875-893.

Coates, Irene. *The Seed Bearers – role of the female in biology and genetics.* Durham: Pentland Press, 1993.

Conn, Sarah A. "The Self-World Connection." *Woman of Power,* Issue 20, Spring 1991: 71-77.

_________. "When Earth Hurts, Who Responds?" in *Ecopsychology: Restoring the Earth, Healing the Mind.,* edited by T. Roszak, M.E. Gomes, & A.D. Kanner. 156-171. San Francisco: Sierra Books, 1995.

Coulson, Sheila. "World's Oldest Ritual Discovered – Worshipped the Python 70,000 years ago," Science Daily, November 30, 2006.

Source: The Research Council of Norway.

Crowley, Vivianne. *Celtic Wisdom: Seasonal Rituals and Festivals*. New York: Sterling, 1998.

Daly, Mary. *Gyn/Ecology: The Metaethics of Radical Feminism*. London: The Women's Press, 1979.

Dames, Michael. *Ireland: a Sacred Journey*. Element Books, 2000.

de Beauvoir, Simone. *The Second Sex*., translated by H.M. Parshley. New York: Knopf, 1953.

Denning, M. et al. *The Magical Philosophy III*. Minnesota: St.Paul, 1974.

De Shong, Betty. *Inanna Lady of Largest Heart: Poems of the Sumerian High Priestess Enheduanna*. Texas: University of Texas Press, 2000.

Devereux, Paul. *Earth Memory: The Holistic Earth Mysteries Approach to Decoding Ancient Sacred Sites*. London: Quantum, 1991.

Dexter, Miriam Robbins. *Whence the Goddesses: A Source Book*. New York: Teacher's College Press, 1990.

______. "The Ferocious and the Erotic: 'Beautiful' Medusa and the Neolithic Bird and Snake." *Journal of Feminist Studies in Religion*, Issue 26.1 (2010): 25-41.

Dimitrov, Vladimir. *Introduction to Fuzziology: Study of Fuzziness of Knowing.* Lulu.com, 2005.

Dimitrov, Vladimir and Hodge, Bob. *Social Fuzziology: Study of Fuzziness of Social Complexity*. Heidelberg New York: Physica-Verlag, 2002.

Downing, Christine. *The Goddess: Mythological Images of the Feminine*. New York: Crossroad, 1984.

Drury, Neville. *The Elements of Shamanism.* Element Books: Dorset, 1989.

Durdin-Robertson, Lawrence. *The Year of the Goddess.* Wellingborough: Aquarian Press, 1990.

Edwards, Carolyn McVickar. *The Storyteller's Goddess Tales of the Goddess and Her Wisdom from Around the World.* New York: HarperCollins, 1991.

Eiseley, Loren. *The Immense Journey.* New York: Vintage Books, 1957.

Eisler, Riane. *The Chalice and the Blade.* San Fransisco: Harper and Row,1987.

Flamiano, Dominic. "A Conversation with Brian Swimme." *Original Blessing,* Nov/Dec 1997: 8 -11.

Fox, Matthew, editor. *Original Blessing, a Creation Spirituality Network Newsletter.* Friends of Creation Spirituality Inc.: Oakland CA, 1990's.

French, Claire. *The Celtic Goddess Great Queen or Demon Witch?* Edinbu rgh: Floris Books, 2001.

Gadon, Elinor W. *The Once and Future Goddess.* Northamptonshire: Aquarian, 1990.

Gebser, Jean. *The Ever-Present Origin,* translated by Noel Barstad with Algis Mickunas. Athens Ohio: Ohio University Press, 1985.

George, Demetra. *Mysteries of the Dark Moon.* San Francisco: HarperCollins, 1992.

Getty, Adele. *Goddess: Mother of Living Nature.* London: Thames and Hudson, 1990.

Gimbutas, Marija. *The Goddesses and Gods of Old Europe.* Berkeley and Los Angeles: University of California Press, 1982.

___________. *The Language of the Goddess.* New York: HarperCollins, 1991.

___________ *The Living Goddesses,* edited and supplemented by Miriam Robbins Dexter. Berkeley and Los Angeles: University of California Press, 1999.

Glendinning, Chellis. *My Name is Chellis and I'm in Recovery from Western Civilization.* Boston: Shambhala Publications, 1994.

Goodenough, Ursula. *The Sacred Depths of Nature.* New York and Oxford: Oxford University Press, 1998.

Grahn, Judy. "From Sacred Blood to the Curse and Beyond." In *The Politics of Women's Spiritualty,* edited by Charlene Spretnak, 265-279. New York: Doubleday, 1982.

_________. *Blood, Bread and Roses: How Menstruation Created the World.* Boston: Beacon Press, 1993.

Gray, Susan. *The Woman's Book of Runes.* New York: Barnes and Noble, 1999.

Greene, Brian. *The Elegant Universe.* London: Vintage, 2000.

Gross, Rita. "The Feminine Principle in Tibetan Vajrayana Buddhism." *The Journal of Transpersonal Psychology.* Vol.16 No.2, 1984: 179 -192.

Halifax, Joan. *Being With Dying.* (CD series) Colorado: Sounds True, 1997.

Harding, M. Esther. *Women's Mysteries, Ancient and Modern.* London: Rider & Company, 1955.

Harman, W. & Sahtouris, E. *Biology Revisioned.* Berkeley: North Atlantic Books, 1998.

Harrison, Jane Ellen. *Themis: A Study of the Social Origins of Greek Religion.* Cambridge: Cambridge University Press, 1912.

____________ *Myths of Greece and Rome.* London: Ernest Benn Ltd., 1927.

_____________ *Prolegomena to the Study of Greek Religion.* New York: Meridian Books, 1957.

Holler, Linda. "Thinking with the Weight of the Earth: Feminist Contributions to an Epistemology of Concreteness." *Hypatia* Vol. 5 No. 1, Spring 1990: 1–22.

Houston, Jean. *The Search for the Beloved: Journeys in Mythology and Sacred Psychology.* Los Angeles: Jeremy P. Tarcher, Inc., 1987.

___________ *The Hero and the Goddess.* New York: Aquarian Press, 1993.

Hwang, Helen Hye-Sook. *The Mago Way: Re-discovering Mago, the Great Goddess of East Asia (Volume 1).* Mago Books, 2015.

__________________ "Goma, the Shaman Ruler of Old Magoist East Asia/Korea, and Her Mythology, in *Goddesses in Myth, History, and Culture.* Lytle Creek: Mago Books, 2018, 252-276.

__________________ "Unveiling an Ancient Sill Korean Testimony to the Mother World: An Introductory Discussion of the Budoji (Epic of The Emblem Capital City), the Principal Text of Magoism" in *S/HE: An International Journal of Goddess Studies,* Vol 1 No.2, 2022: 4-70.

Johnson, Robert A. *She: Understanding Feminine Psychology.* New York: Harper and Row, 1977.

Jones, Kathy. *Priestess of Avalon, Priestess of the Goddess.* Glastonbury: Ariadne Publications, 2006.

Kremer, Jürgen. "Post-modern Shamanism and the Evolution of Consciousness" paper delivered at the International Transpersonal Association Conference, Prague, June 20-25, 1992. The audio tape no longer seems to be available.

Lao Tzu. *The Way of Life*, translated by Witter Bynner. New York: Capricorn Books, 1962.

Larousse Encyclopedia of Mythology, London: Hamlyn Publishing Group Ltd., 1968.

Lederer, Wolfgang. *The Fear of Women.* New York: Harcourt Brace Jovanovich Inc., 1968.

Livingstone, Glenys. *The Female Metaphor – Virgin, Mother, Crone – of the Dynamic Cosmological Unfolding: Her Embodiment in Seasonal Ritual as Catalyst for Personal and Cultural Change.* Ph.D. thesis, University of Western Sydney, 2002. http://www.academia.edu/27860395/The_female_metaphor_-_virgin_mother_crone_of_the_dynamic_cosmological_unfolding_her_embodiment_in_seasonal_ritual_as_a_catalyst_for_personal_and_cultural_change.

__________. *PaGaian Cosmology: Re-inventing Earth-based Goddess Religion.* Nebraska: iUniverse, 2005.

__________. "Gaia: Dynamic, Diverse, Source of Being." In *Goddesses in World Culture*, edited by Patricia Monaghan, 143-154. Praeger, 2011.

Lorde, Audre. "Uses of the Erotic." In *Weaving the Visions: New Patterns in Feminist Spirituality,* edited by Judith Plaskow & Carol Christ, 208-213. New York: HarperCollins, 1989.

Macy, Joanna and Brown, Molly Young. *Coming Back to Life.* Gabriola Island, Canada: New Society Publishers, 1998.

Matthews, Caitlin. *The Celtic Spirit.* London: Hodder and Stoughton, 2000.

Matthews, Caitlin and John. *The Western Way.* London: Penguin, 1994.

McHardy, Stuart. "Bride of Scotland." In *Brigit: Sun of Womanhood,* edited by Patricia Monaghanand Michael McDermott, 49-58. Nevada: Goddess Ink, Ltd., 2013.

McLean, Adam. *The Four Fire Festivals.* Edinburgh: Megalithic Research Publications, 1979.

__________. *The Triple Goddess.* Grand Rapids Michigan: Phanes Press, 1989.

Merleau-Ponty, Maurice. *The Invisible and the Invisible,* translated by Alphonso Lingis. Evanston, Illinois: Northwestern University Press, 1968.

Monaghan, Patricia. *O Mother Sun! A New View of the Cosmic Feminine.* Freedom CA: Crossing Press, 1994. *

____________, editor. *Goddesses in World Culture*, Vol 2. Praeger, 2011.

______________ and McDermott, Michael, editors. *Brigit: Sun of Womanhood.* Nevada: Goddess Ink, Ltd., 2013.

Morgan, Robin. "The Network of the Imaginary Mother." In *Lady of the Beasts,* 63-88. New York: Random House, 1976.

Murphy, Susan. *upside-down zen: a direct path into reality.* Melbourne: Lothian, 2004.

Neumann, Erich. *The Great Mother.* Princeton: Princeton University Press, 1974.

Nichols, Mike. "The First Harvest." *Pagan Alliance Newsletter,* NSW Australia, Lughnasad 2000, Vol. 4, No. 7: 1.

Noble, Vicki. *The Double Goddess: Women Sharing Power.* Vermont: Bear & Company, 2003.

Orenstein, Gloria Feman. *The Reflowering of the Goddess.* New York: Pergamon Press, 1990.

Orr, Emma Restall. *Spirits of the Sacred Grove.* London: Thorsons, 1998.

Pagels, Elaine. *Beyond Belief: the Secret Gospel of Thomas.* New York: Random House, 2003.

Pirtle, Sarah. "A Cosmology of Peace." *EarthLight,* Vol. 15, No. 1, Fall 2005: 10-11.

Pollack, Rachel. *The Body of the Goddess.* Brisbane: Element Books, 1997.

Raphael, Melissa. *Thealogy and Embodiment: the Post-Patriarchal Reconstruction of Female Sexuality.* Sheffield: Sheffield Press, 1996.

Reinach, Salomon. *Orpheus.* New York: Horace Liveright, Inc., 1930.

Reis, Patricia. "The Dark Goddess." *Woman of Power,* Issue 8, Winter 1988: 24 –27, 82.

Rich, Adrienne. *Of Woman Born.* New York: Bantam, 1977.

Rigoglioso, Marguerite. *The Cult of Divine Birth in Ancient Greece.* New York:Palgrave Macmillan, 2009.

Ross, Nancy Wilson. *Three Ways of Asian Wisdom.* New York: Simon & Schuster, 1966.

Sahtouris, Elisabet. *Earthdance: Living Systems in Evolution.* Lincoln Nebraska: :iUniversity Press, 2000.

Sewall, Laura. "Earth, Eros, Sky." *Earthlight*, Winter 2000: 22-23 and 25.

Shuttle, Penelope and Redgrove, Peter. *The Wise Wound: Menstruation and Everywoman.* London: Paladin Books, 1986.

Sikie, Elisabeth. *Patterns of Invocation in Neolithic Art: Reclaiming the Indigenous Mind*, a paper written 2004-2005, no longer available.

Sjöö, Monica and Mor, Barbara. *The Great Cosmic Mother: Rediscovering the Religion of the Earth.* San Francisco: Harper and Row, 1987.

Solomon. Annabelle. *The Wheel of the Year: Seasons of the Soul in Quilts.* Winmalee NSW: Pentacle Books, 1997.

Spretnak, Charlene, ed. *The Politics of Women's Spirituality.* New York: Doubleday, 1982.

___________. "Mythic Heras as Models of Strength and Wisdom." In *The Politics of Women's Spirituality*, edited by Charlene Spretnak, 87-90. New York: Doubleday, 1982.

_____________. "Gaian Spirituality." *Woman of Power*, Issue 20, Spring 1991: 10 -17.

_____________. *Lost Goddesses of Early Greece: a Collection of Pre-Hellenic Myths.* Boston: Beacon Press, 1992.

_____________. *States of Grace: The Recovery of Meaning in the Postmodern Age.* San Francisco: HarperCollins, 1993.

_____________. *The Resurgence of the Real: Body, Nature and Place in a Hypermodern World.* New York: Routledge, 1999.

Starhawk. *Truth or Dare: Encounters with Power, Authority, and Mystery*. San Francisco: Harper and Row, 1990.

________ *The Spiral Dance: A Rebirth of the Ancient Religion of the Great Goddess*. New York: Harper and Row, 1999.

_______. *The Earth Path*. San Francisco: HarperCollins, 2004.

Stockton, Eugene, ed. *Blue Mountains Dreaming*. Winmalee: Three Sisters Productions, 1996.

Stone, Merlin. *When God was a Woman*. London: Harvest/HBJ, 1978.

__________. *Ancient Mirrors of Womanhood Vol 1*, Boston: Beacon Press, 1984.

Swimme, Brian. *The Universe is a Green Dragon*. Santa Fe: Bear & Co., 1984.

___________. *Canticle to the Cosmos*. DVD series. CA: Tides Foundation, 1990.

__________ *Canticle to the Cosmos: Study Guide*. Boulder CO: Sounds True Audio, 1990.

__________. *The Hidden Heart of the Cosmos*. New York: Orbis, 1996.

______________ *The Hidden Heart of the Cosmos*. Video: Center for the Story of the Universe, 1996.

__________. *The Earth's Imagination*. DVD series, 1998.

__________. *The Powers of the Universe*, DVD series, 2005.

Swimme, Brian and Berry, Thomas. *The Universe Story: From the Primordial Flaring Forth to the Ecozoic Era*. New York: HarperCollins, 1992.

Tacey, David. "Spirit and Place." *EarthSong Journal*, issue 1, Spring 2004: 7-10 and 32-35.

Taylor, Dena. *Red Flower: Rethinking Menstruation.* Freedom CA: Crossing Press, 1988.

Tedlock, Dennis, and Barbara Tedlock, eds. *Teachings from the American Earth.* New York: Liveright, 1975.

Thadani, Gita. *Sakhiyani: Lesbian Desire in Ancient and Modern India.* London: Cassell, 1996.

Toulson, Shirley. *The Celtic Year.* Dorset: Element Books Ltd., 1993.

Vernadsky, Vladimir. *The Biosphere.* London: Synergetic Press, 1986.

Walker, Barbara. *The Woman's Encyclopedia of Myths and Secrets.* San Francisco: Harper and Row, 1983.

_____________. *The Crone: Woman of Age, Wisdom and Power.* New York: HarperCollins, 1988.

Warner, Marina. *Alone of All Her Sex.* New York: Alfred Knopf, 1976.

Webster's Third International Dictionary of the English Language Unabridged. Encyclopaedia Britannica, 1986.

Wilber, Ken. *A Brief History of Everything.* Massachusetts: Shambhala, 1996.

INDEX

Lady of the Forge - permenopause
smith - transformation in fire

'I wake up burning, my skin ~~glowing~~ glistening like molten metal: liquid fire dripping from my forehead, and my half-dreaming moan is guttural.

Printed in Great Britain
by Amazon